Underwater Archaeology of

Dwarka and Somnath

(1997-2002)

Underwater Archaeology of

Dwarka and Somnath

(1997-2002)

A.S. Gaur, Sundaresh and K.H. Vora

ARYAN BOOKS INTERNATIONAL
New Delhi

NATIONAL INSTITUTE OF OCEANOGRAPHY
Goa

UNDERWATER ARCHAEOLOGY OF DWARKA AND SOMNATH
(1997-2002)

ISBN-10: 81-7305-349-9
ISBN-13: 978-81-7305-349-8

First Published in **2008** by:
ARYAN BOOKS INTERNATIONAL
Pooja Apartments, 4B, Ansari Road, New Delhi-110 002 (India)
Tel.: 23287589, 23255799; Fax: 91-11-23270385; E-mail: aryanbooks@vsnl.com

in association with

NATIONAL INSTITUTE OF OCEANOGRAPHY
Dona Paula, Goa-403 004

Computer Typeset and Printed in India at
ABI Prints & Publishing Co., New Delhi

Dedicated

to

The Jagat Mandir, Dwarkadhish

Foreword

National Institute of Oceanography, Goa initiated marine archaeological research in India by establishing Marine Archaeology Centre in 1983. Dwarka was the site that was studied first. It had attracted historian and archaeologists since early years of the last century. Marine archaeological explorations have attracted the media and the common man ever since the Dwarka studies because of Dwarka's links to ancient texts like the *Mahabharata, Harivamsa* and *Vishnu Purana*.

The present book summarizes results of the underwater explorations carried at Dwarka and Somnath between 1998 and 2001 as a joint project of Department of Ocean Development (now Ministry of Earth Sciences) and Council of Scientific and Industrial Research. The main feature of this exploration (Phase II) was identification of objects made from stone. A number of stone anchors were recovered from different depths during the exploration of Dwarka and Somnath sites. These finds indicate that intense maritime activities were undertaken in the Okhamandal region in the past. Underwater exploration at Somnath, which is described in *Mahabharata* as Prabhas, yielded a large quantity of stone anchors many of which are ring-shaped.

This book discusses extensively the date and identification of anchors and other objects recovered from underwater exploration. The book also addresses some of the controversies that arose during early stage of the exploration based on additional facts and findings collected. It is hoped that the book will shed more light on the underwater heritage of Dwarka for the benefit of scholars of history and archaeology, and students.

Satish R. Shetye
Director
National Institute of Oceanography

The Team - *Sitting*: Sundaresh, K.H. Vora, A.S. Gaur;
Standing: S.B. Chitari, S.N. Bandodkar, Sila Tripati

Preface

Indian subcontinent is the land of living civilization through the ages as demonstrated by several recent archaeological investigations carried out across the subcontinent. Fields like agriculture, metallurgy, education, religion, town planning and water harvesting are some of the areas where the Indian subcontinent reached the culmination of developmental excellence since the 3rd millennium BC. Similarly, on the external front, the subcontinent had overseas trade relations with its contemporary civilizations like the Mesopotamian, the Elam and other cultures of the Persian Gulf such as Barbar and Umm an Nar of the west. The advancement in maritime activities of the Bronze Age civilization of the subcontinent is demonstrated through archaeological artefacts, such as fishhooks, boat motifs on seals, painting on potsherds, marine shell artefacts, fish bones, etc. It is this legacy of maritime activities, that we have studied during the last two-and-a-half decades and accumulated information that needed to be disseminated among scholars and students in the form of this book. The development of new technologies for underwater explorations like scuba diving, marine geophysical equipment, underwater metal detectors, underwater photography, dating techniques, etc. have facilitated new field of science of marine archaeology to emerge as follow through of global development. In India, S.R. Rao initiated marine archaeological investigations in 1981, and Dwarka on the west coast of India was an automatic choice to trace the authenticity of the popular tradition of the submergence of town in the sea and its relation to study the sea level history during the historical times.

Dwarka, an attraction to Hindu pilgrims, archaeologists, historians, and oceanographers, has been responsible for the establishment of the Marine Archaeology Centre at the National Institute of Oceanography, Goa. Since then, a number of expeditions of marine archaeological investigations were organized in Dwarka along with Bet Dwarka and Somnath and a large amount of data were collected to demonstrate the rich maritime traditions along the Indian coast.

The findings of Dwarka have evoked tremendous interest in scholars and general public. A large number of research papers and a book entitled *The Lost City of Dvaraka* have earlier been published by S.R. Rao, dealing with the work carried out between 1983 and 1994 emphasizing his hypothesis of correlating the findings with the description of the site mentioned in various ancient Indian literature, including the *Mahabharata*, the *Vishnu Purana* and the *Harvamsa Purana*. However, simultaneously, Rao's work on Dwarka has been commented upon by various scholars, especially for the dating of the underwater structures (rather, lack of scientific dates from underwater artefacts) and correlation with the *Mahabharata*. These concerns helped us to identify what more can be done and has to be done. Consequently, a scientific proposal was submitted to the Department of Ocean Development (now the Ministry of Earth Sciences) as a second phase.

The second phase of the underwater explorations commenced in 1998 and continued till 2001. The findings from this phase provided platform for a radical change in the hypothesis based on facts and findings. The present book comprises the underwater archaeological explorations undertaken during the second phase at Dwarka and Somnath between 1998-2001. The archaeological findings from Dwarka have been compared with the findings from various sites along the Saurashtra coast to provide relative information. The majority of the findings presented in this book have already been published in various national and international journals; however, they have been synthesized with uniform analysis and interpretation here.

The underwater and coastal explorations over the two decades along the Indian coast brought to light a variety of artefacts and structures related to ancient maritime activities. The type of the antiquities recovered from the sea bed, is indicative of the involvement of ancient man in maritime activities. The most important and direct evidence of maritime activities is an anchor left along the coast by ancient sailors many centuries before. The variation in sizes and type of stone anchors indicate the experiments that have been carried out to meet the requirement of anchoring system depending upon the size of boat, type of sea bed and nature of harbour. Our explorations clearly indicate that ancient navigators used to carry a large number of stone anchors on board to deploy in different sea bed conditions.

The book has been divided into two major sections, each devoted to Dwarka and Somnath, respectively. Section I deals with underwater explorations at Dwarka. Further, this section is divided into 5 chapters. The Introduction comprises scope and objective in Chapter 1, while Chapter 2 is devoted to the methodology of the work and equipment used during the underwater investigations at Dwarka and Somnath. The next chapter describes the results of the explorations at Dwarka and discusses various hypotheses on the identification of submerged stone structures. Chapter 4 exclusively provides the details of the stone anchors recovered during the explorations at Dwarka. Chapter 5 deals with a few additional sites of Okhamandal, such as Aramda and Positra. Chapter 6 provides concluding remarks on the findings of Dwarka. Section II is exclusively devoted to the explorations at Somnath. In this case also, the introductory chapter deals with details of the site, scope, objective and the background information (literary, geological and archaeological) of the site. The underwater findings have been described in Chapter 2. The last chapter of this book is a concluding remark on Somnath explorations. A combined reference and bibliography have been provided at the end. The book has been fully illustrated with underwater and surface photographs, drawings and tables. It is hoped that the book will provide valuable facts on the underwater archaeology of Dwarka and Somnath for readers and scholars to draw their own conclusions.

A.S. Gaur
Sundaresh
K.H. Vora

Contents

Acknowledgements

We are extremely beholden to Dr. S.R. Rao, father of Marine Archaeology in India, for initiating research in the field of marine archaeology at Dwarka, which culminated with the establishment of the Marine Archaeology Centre at the National Institute of Oceanography, Goa. We thank the Director of the National Institute of Oceanography, Goa, for his continuous support for the marine archaeological studies along the Indian coast. The Department of Ocean Development (now Earth Science Commission) and Council of Scientific and Industrial Research, New Delhi funded this project, for which we are thankful. Thanks are also due to the project review committee members, Shri R.S. Bisht, former Joint Director General, Archaeological Survey of India; Prof. S.N. Rajaguru, former Joint Director, Deccan College, Pune; Prof. V.H. Sonawane, Department of Archaeology, M.S. University, Baroda; Dr. S.P. Gupta, Chairman of Society of Indian Archaeology, New Delhi; Dr. K. Rajan, Department of Archaeology, Tamil University, Thanjavur; and Dr. A.G. Untawale, former Dy. Director, NIO for their recommendations and suggestions on the underwater findings from Dwarka time to time.

Late Shri Pushkarbhai Gokani, a well known philanthropist from Dwarka, was a constant source of inspiration and he supported this work by discussing various aspects during our visits to his home at Dwarka. We also thank delegates of the 51st *Gyan Satra* of Gujarat Itihas Parishad held at Dwarka in October 2003 for enlightening us on the various aspects of the history of Dwarka.

Our fieldwork would have not been possible without the help and support rendered by Shri S.N. Bandodker, who was the dive master for underwater operation at Dwarka and Somnath besides photographing underwater findings and handling of equipment on board. Dr. Sila Tripati has continuously dived with us at Dwarka and Somnath; Shri S.B. Chitari, who rendered invaluable assistance in variety of ways not only on the field and but also in the laboratory; Shri G.R. Desai assisted during various activities on field; Shri K.L. Naik helped in profiling of the sea bed with echo sounder. We thank them all. Shri R. Uchil, Shri P. Pawaskar and Shri K.G. Chitari helped in preparation of computer illustrations. Shri Umesh Sirsat, Shri Shaikh

Ali Karim and Shri M. Mochhemadkar did photography at the laboratory. We thank them all for all their help and support.

Shri Abbas Suleman, owner of the boats *Gangasagar* and *Lakshminarayan,* is especially remembered for his contribution beyond the code of contract in organizing the expeditions and taking personal care of our equipment and all the team members. We sincerely thank him. Last but not the least, we thank our stakeholders, i.e., scholars, experts, students and common men who regularly queried us on the different aspects of our findings and inspired us to prepare this book.

Abbreviations

AJA = American Journal of Archaeology
ASI = Archaeological Survey of India
AT = Aramda Triangular anchor
BAIMA = Bulletin of Australian Institute of Maritime Archaeology
BC = Before Christ
BP = Before Present
CS = Current Science
CSIR = Council of Scientific and Industrial Research
DG = Dwarka Grapnel anchor
DOD = Department of Ocean Development
DST = Department of Science and Technology
DR = Dwarka Ring stone anchor
DT = Dwarka Triangular anchor
GMAI = Glimpses of Marine Archaeology in India
GPS = Global Positioning System
IAR = Indian Archaeology – A Review
IJNA = International Journal of Nautical Archaeology
INSA = Indian National Science Academy
JGSI = Journal of Geological Society of India
JIOA = Journal of Indian Ocean Archaeology
JMA = Journal of Marine Archaeology
JRAS = Journal of the Royal Asiatic Society
MAIOC = Marine Archaeology of Indian Ocean Countries
MASI = Memoir of Archaeological Survey of India
ME = Man and Environment
MM = Mariner's Mirror
NIO = National Institute of Oceanography, Goa

OBM = Outboard Machine
QR = Quaternary Research
RASB = Royal Asiatic Society of Bangladesh
RURIMA = The Role of Universities and Research Institutes in Marine Archaeology
SCUBA = Self-Contained Underwater Breathing Apparatus
SDDE = Surface Demand Diving Equipment
SG = Somnath Grapnel anchor
SR = Somnath Ring stone anchor
ST = Somnath Triangular anchor

List of Illustrations

FIGURES

PLATES

Section I
DWARKA

1
Introduction

The study of maritime activities of ancient India has been a fascinating aspect of our history and was initiated as early as in the beginning of the 20th century (Mookerji, 1912). Initially, it was largely based on data from the ancient Indian literature such as *Vedas, Puranas, Upanishads,* epics and other Buddhist and Jain literature. The foreign accounts such as the *Periplus of the Erythraean Sea,* Ptolemy's *Geography* and the *Indica* of Megasthenese provide clues on the maritime supremacy of India during the Early Historical Period. Besides literature, cave paintings and rock engravings on various monuments also added to the knowledge as the primary sources of study. However, hardcore archaeological data on the maritime activities started emerging only in the second-half of the 20th century AD from the excavations at various Harappan sites in Gujarat (Fig. 1) and is continuing till date. The establishment of the Marine Archaeology Centre at the National Institute Oceanography, Goa in the last quarter of the 20th century AD provided much needed impetus to research in the field of marine archaeology in India.

The marine archaeological investigations around the world have revealed several ancient habitations – those that have been submerged in the sea due to perhaps sea level fluctuation, coastal erosion and tectonic disturbances. The Marine Archaeology Centre of the National Institute of Oceanography, Goa has investigated several places along the Indian coast, namely, Dwarka, Bet Dwarka, Somnath, Vijaydurg, Sindhudurg, Goa, Lakshadweep, Poompuhar and Mahabali-puram during the last two-and-a-half decades. The findings are not only interesting but also lead to directions for further research in multi-disciplinary areas.

Initially, marine archaeological explorations off Dwarka, Bet Dwarka and Somnath brought to light very encouraging results, which led us to expand and explore it extensively. A large portion of the historical period (2200 years BP) habitation has been submerged in the sea off Bet Dwarka Island. Several stone structures have been located in shallow waters off Dwarka and they have been described as walls, bastions and pillar and suggested that these are the remains of ancient habitation (Rao, 1990). Besides stone structures, a large number of different types of stone anchors have been reported from the intertidal zone to 16 m water depths off Dwarka. Similarly, a large number of stone anchors have also been found in Somnath water. These

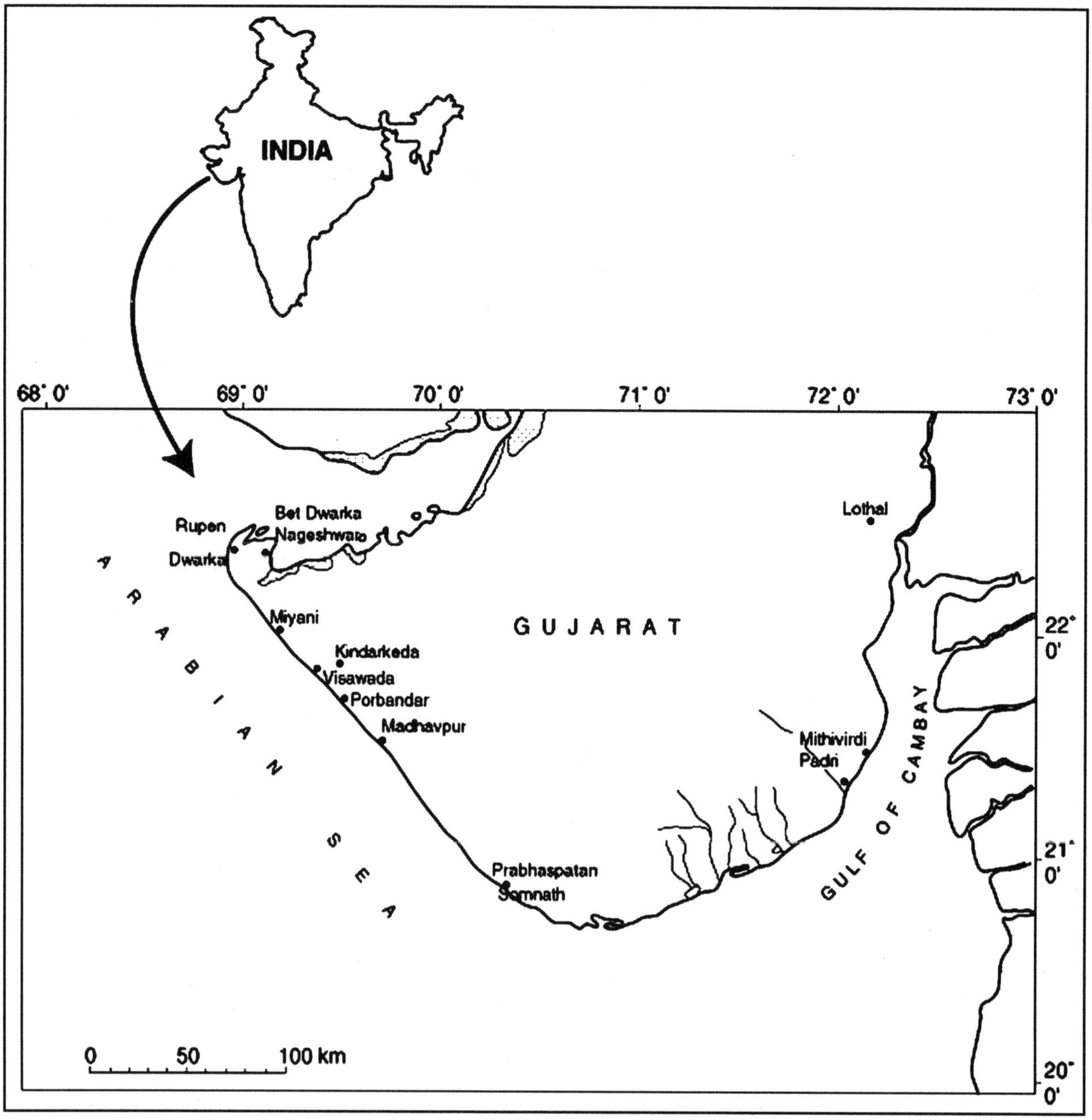

Fig. 1. Location map of coastal archaeological sites along the Saurashtra Coast.

evidences suggest that these were the busiest ports during the historical times on the Saurashtra coast.

The explorations off Dwarka by the Marine Archaeology Centre of the National Institute of Oceanography, Goa, were initiated in 1983 with a small funding from the Indian National Science Academy (INSA), and later small scale funding was continued with other agencies like the Archaeological Survey of India (ASI), the Department of Science and Technology (DST), etc. A proposal for further survey was submitted to the Department of Ocean Development (DOD) in 1994 and ultimately the project was sanctioned jointly by the Council of Scientific and Industrial Research (CSIR) and DOD for three years. The project was initiated in 1998-99 season with the following objectives for further explorations off Dwarka and Somnath.

1. To trace seaward extension of the submerged remains of ancient Dwarka.
2. To search and expose the existing stone structures with the help of airlift operations and document all the structural remains during the first phase of explorations in Dwarka waters.
3. Onshore and intertidal zone exploration of surrounding sites to trace similar structures to correlate the submerged structures of Dwarka.
4. To carry out detailed onshore and offshore explorations at Somnath to find out the archaeological remains to correlate with Dwarka, if any.

As a part of project requirement, extensive onshore explorations were carried out in Okhamandal area (Fig. 2) and surroundings of Somnath to correlate the underwater findings with onshore remains. Offshore explorations at Somnath revealed similar remains like those found in Dwarka. To collect data from offshore diving is an important tool. The standard methods of diving like Surface Demand Diving Equipment (SDDE) and Self-Contained Underwater Breathing Apparatus (SCUBA) system play an important role to meet the objectives and the same were used during 1998-2001 explorations at Dwarka and Somnath. Critical underwater documentation techniques such as drawing, photography and videography were also implemented.

1.1 BACKGROUND INFORMATION

The explorations have been carried out at two places, namely, Dwarka, and Somnath which are supposed to be contemporary but have different geological, geographical and archaeological background. A brief description of the same for Dwarka will follow here while for Somnath, it is given in Section II.

1.2 GEOLOGY OF OKHAMANDAL

Sediment formation of the area is mainly characterized by different types of clays, sandstone and limestone deposits, which have been described as the Gaj and Dwarka beds (Fedden, 1884) and dated to Miocene for Gaj beds and Mio-Pliocene to Pleistocene for Dwarka beds. Some other works (Jain, 1990, 1997; Jain & Agrawal, 1992) have suggested threefold classification of the Dwarka formation based on lithological variations. Accordingly, the Dwarka formation was subdivided into three members, viz., lower member, middle member and upper member ranging in age from Middle Miocene to Pliocene. The Gaj formation was assigned a Lower to Middle Miocene age. Sakhawala and Shringarpure (1990), however, suggested Pleistocene age for the Dwarka rocks based on their study in Bet Dwarka Island. Recently, Bhatt (2000) has prepared a geological map of this area and suggested two major faults at Okha Rann and Bet Dwarka. He has classified Gaj formation into two members: (1) Ashapura Clay Member characterized by grey and variegated clays, marls with some gypsum bands, and (2) Ranjitpur Limestone Member which is yellow and brown fossiliferous and the age to assigned them is Lower to Middle Miocene. Dwarka formation was classified into three members: (i) Positra Limestone characterized by bioclastic and coralline limestone with few dolomitic bands and dated to Middle Miocene, (ii) Shankhodhar Sand-clay Member consisting of grey and yellow sandy clays and sandstone dated to upper miocene, and (iii) Kalyanpur Limestone Member characterized by pinkish to light brown coloured re-crystallized fossiliferous limestone dated to Lower Pliocene. The third important formation has been named as Chaya formation which is also classified into two members: (i) Okha-Shell Limestone characterized by off-white coloured medium to very coarse grained highly fossiliferous porous limestones assigned to Middle to Late Pleistocene age, and (ii) Armada Reef Member consisting of coral reef formation and a date assigned to Late Pleistocene to Holocene period.

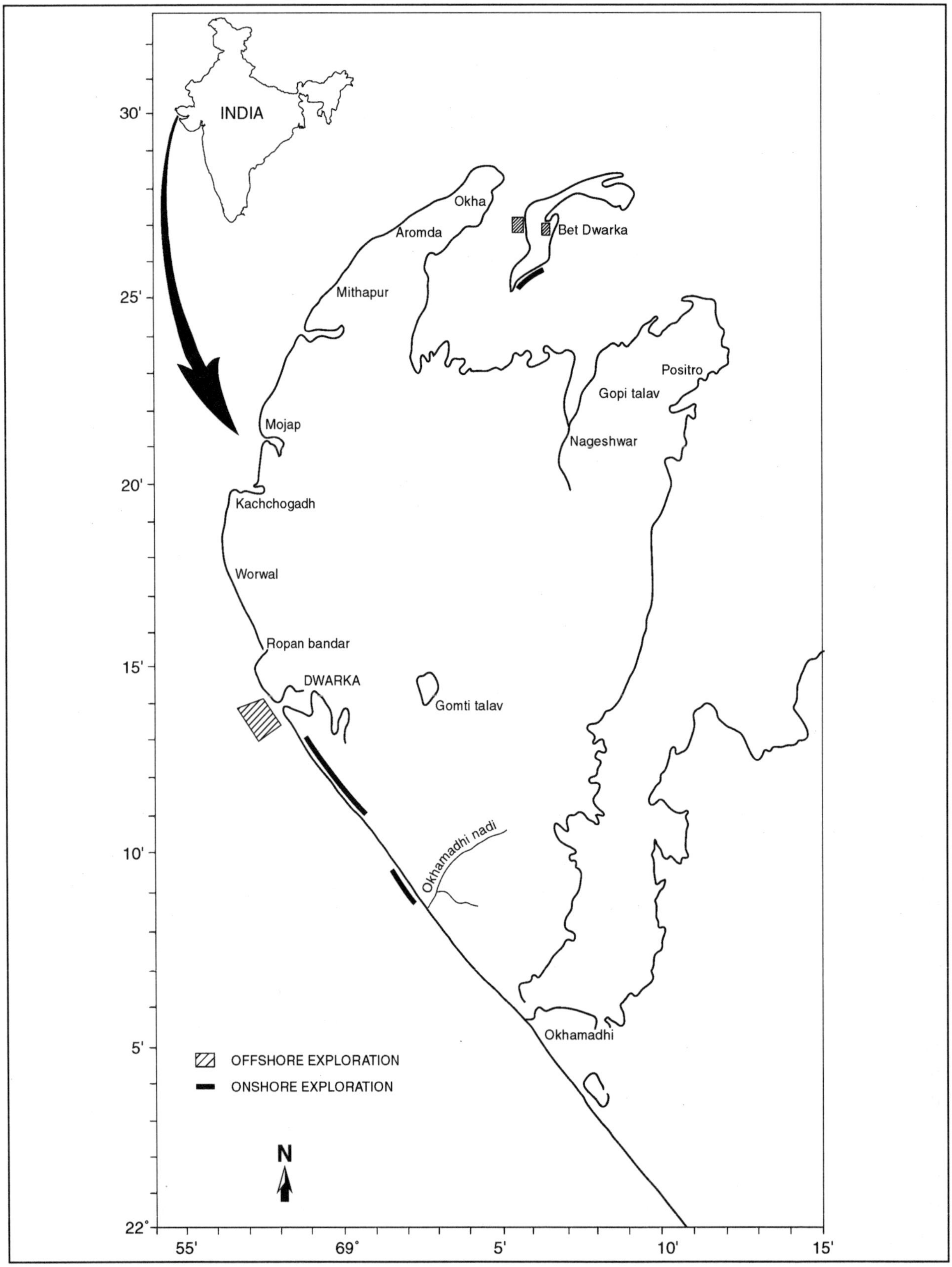

Fig. 2. Map of Okhamandal region with location of Dwarka.

1.3 DWARKA IN LITERATURE AND TRADITION

Dwarka, one of the most important religious places for Hindus, has also attracted tourists from all over the world due to its fabulous architectural planning of the Dwarkadhish temple (Pl. 1). It is situated on the extreme west coast of the Indian territory. The place has association with Lord Krishna who is believed to be the founder of this town by reclaiming 12 *yojana* land from the sea (*Vishnu Purana,* 5.23.13). The town consisted of beautiful gardens, deep motes, several ponds and palaces (*Vishnu Purana,* 5.23.14). The glorious town of Dwarka was submerged in the sea just after the death of the Lord Krishna (*Mahabharata, Mausala Parva,* 7.41-42). As Dwarka has the historical importance and association with the *Mahabharata,* the site has attracted several archaeologists and historians besides scientists.

There have been various references on the location of Dwarka in different ancient Sanskrit texts (Shastri, 1941). For instance, *Adi* and *Sabha Parvas* of the *Mahabharata* referred to a Dvaraka near the hill Raivataka, and no reference at all was made to the city being near the sea or its later drowning by natural calamity. In the *Mausala Parva,* Dwarka was close to the sea and later destroyed by the sea. The same view seems to have been held by the *Puranas:* the *Vayu* and the *Vishnu* agreeing with the *Adi* and *Sabha Parvas* and the *Bhagavata, Harivamsa* and others following the *Mausala* (Sankalia, 1966).

Bombay Gazetteer has noted a tradition of Dwarka, i.e.,

> The earliest known figure is one Sukkur Belim, 'a Syrian adventurer', who established his hold in Okhamandal. A popular tradition says that Dwarka was swallowed up by the sea during his reign. This did not deter the ruler who established himself at Gorinja, a village 9 km to the southeast of modern Dwarka (Gazetteer of Bombay, 1884).

The Kshatrapas ruled the western India including Gujarat and Saurashtra for the first four centuries of the Christian era. A 2nd century AD inscription of the Kshatrapa regime found at Mulwasar, near Dwarka, indicates Okhamandal was directly ruled by the Western Kshatrapas from the very beginning of the Christian era. The next important rulers were Guptas. A large number of Gupta coins of Saurashtra fabric are found at Dwarka and Pindara (Ansari and Mate, 1966).

Dwarka has been an attraction for historians since the beginning of the 20th century and several literary references have been cited to locate Dwarka of the Mahabharata period. Pargiter (1904) was first to suggest that Dwarka was located near the Raivataka mountain and elsewhere also he has mentioned that it was constructed on the remains of Kusasthali (Pargiter, 1922). Pusalkar (1943 & 1945) accepted that modern Dwarka was the original Dwarka of the Mahabharata period. The earliest epigraphical reference about Dwarka came from Palitana copper-plate dated to AD 574 (*EI,* XI). The inscription refers to Dwarka as the capital of the western coast of Saurashtra and still more important, states that Sri Krishna lived here. This reference suggests that as early as the 6th century AD, Dwarka had come to be associated with Sri Krishna and the Yadavas. The establishment of one of the four of his *pithas* at Dwarka by Shankaracharya in the beginning of the 9th century AD attests the great religious sanctity of the place must have attained by the 8th century AD. The Pratihara Emperor Nagabhata II of Kanauj is supposed to have undertaken a pilgrimage of Saurashtra in the early 9th century AD. He visited Dwarka as well as Pindataraka kshetra. The other noteworthy historical reference is to be dated around AD 1500. By this time, Abul Fath Khan, also known as Mahmud Beghra, invaded the Okhamandal area and despoiled the entire area including Dwarka. This mission was ostensibly undertaken in order to suppress the pirates in this area. By 1592 another expedition organized by the Gujarat Governors of the Mughals laid the country low, only to be recovered by the turbulent Vaghers within a few years. It seems that till the beginning of the 19th century

Okhamandal maintained its independence status, often defeated but never completely subjugated. As the numismatic evidence indicates, coins of the Gujarat Sultans were current in this region. But due to its inhospitable nature, the administration of the Sultans or the Mughals was never on a very secure basis. Sankalia (1966) has strongly argued in favour of modern Dwarka as the original Dwarka of the Mahabharata period. Dwarka has also been described as Baraca by the author of the *Periplus of Erythraean Sea* (Schoff, 1974:174).

There have been several other claims for Dwarka in the Saurashtra such as Junagadh or Girinagara, Mula-Dwarka near Prabhash Patan and another Mula-Dwarka between Porbandar and Miyani. However, no serious research has been conducted other than the Dwarka in Okhamandal region.

Recently, Iyangar and Radhakrishnan (2005) have propounded a hypothesis on the location of Dwarka and Prabhasa Patan and predicted that an arm of sea of the Gulf of Khambhat was extended a few hundred km up to southern Rajasthan, and that Saurashtra was an island. Dwarka and Prabhasa Patan must have existed on the eastern side of this channel near the present location of Patan. His hypothesis is based on the study of ancient literature and geological features. However, it requires a critical relook as a number of Harappan sites are located within the said channel and several Mesolithic sites including one well known site at Langhnaj also fall in the channel area whereas the date of Dwarka was suggested much later (3500 yrs BP).

1.4 PREVIOUS ARCHAEOLOGICAL RESEARCH

Two systematic excavations have been carried out by two different respected organizations such as Deccan College, Pune under the direction of H.D. Sankalia in 1963, and Archaeological Survey of India under the direction of S.R. Rao in 1979-80. The brief results of both the excavations are provided below.

1.4.1 Excavation by Deccan College, Pune

The first systematic land excavation at Dwarka was conducted by the Deccan College, Pune in 1963 and dated the earliest occupation of Dwarka to the 1st-2nd century BC/AD. The excavation revealed the existence of four periods (Ansari and Mate, 1966).

Period I is represented by a 5-m thick deposit of sea sand. Besides potsherds of painted pottery, this gave fragments of terracotta balls and stoppers, shells, shell bangles and also a piece of iron. The layer has been dated to the 1st-2nd century BC. The date is assigned mainly on its being anterior to the layer containing Red Polished Ware.

Period II revealed the earliest habitation. The association of the Red Polished Ware and more especially the amphorae in the layers forming this period indicate its date as being the first four centuries of the Christian era.

Period III gives evidence of successive superimposed structures. These are apparently parts of a temple. They bear a close resemblance to the mouldings of the 7th-8th century AD temples in this area.

Period IV represented by polychrome glass bangles, glazed ware and coins of Gujarat Sultans suggest a date around the 10th century AD.

They concluded that due to limited excavation it may be difficult to reconstruct the history of Dwarka and further work may reveal an earlier settlement (Sankalia, 1966).

A few sites, namely, Pindara about 30 km northeast of Dwarka and Suvarntirth about 10 km north of Dwarka, were excavated and both sites have more or less confirmed the chronological sequence as observed at Dwarka. A thick sturdy, Black-on-Red Ware resembling the Kshatrapa pottery from Rangmahal and also from various sites within Saurashtra itself was recovered from both the sites along with the Red Polished Ware and sherds of amphorae (Sankalia, 1966).

1.4.2 Excavation by the Archaeological Survey of India in 1979

Subsequently, a team from the Archaeological Survey of India had demolished a modern building in the forecourt of Dwarkadhish temple in 1979-80, which revealed the remains of a Vishnu temple of the 9th century AD. Further landward excavation revealed two more phases of the temple, datable to the 1st century BC/AD (Rao, 1990). Later, in 1979-80, the Archaeological Survey of India led by S.R. Rao exposed four trenches in the forecourt of Dwarkadhish temple and not very far from the already excavated site by Deccan College, Pune. From Trench DWK-I, layers 1 to 3 are of the Medieval Period and layer 4 has Red Polished Ware of the Early Historical Period. From layer 6 and 7 no artefact is reported while layer 8 has copper coins with Red Polished Ware. According to the excavators, layer 9 contains mixed material of Red Polished Ware and Lustrous Red Ware and layer 10 has exclusive Lustrous Red Ware datable to the mid-2nd millennium BC. The trench DWK-2 revealed a huge structure, which has been described as a secret passage and compared with the description mentioned in the *Harivamsa, Vishnu* and the *Bhagawata Puranas*. No stratigraphy is suggested as the excavator feels it may be filling up. In Trench DWK-3, layers 1-4 two structural phases were noticed which have been assigned to the 15th century AD and the Early Medieval Period pottery in layer 5. Layer 6 and 7 are consisted of fine sand with intervening patches of ash. Red Polished Ware occurs in layer 7 whereas layer 8 has Black and Red Ware which the excavator feels is different from Rangmahal of the Early Historic Period. The excavator claims that layer 9 yielded Lustrous Red Ware and layer 10 revealed the late Harappan pottery. In conclusion of this trench, the excavator says:

> The first settlement at Dwarka was made during the Late Harappan and Post-Harappan periods but the township might have been a little away from the present site. A storm wave, which destroyed the earlier township was responsible for the deposition of the habitational debris here and sending it with sand. As heavy stones were used for building, the entire houses could not be easily carried over long distances by tides; hence, only lighter material like pottery, shells and metal objects were encountered in tidal debris. Sometime after the abandon of first Dwarka in the later half of the 2nd millennium BC, a new folk using the thin Black on Red Painted Ware occupied the site in the first half of the first millennium BC (IAR, 1979-80).

Thus, the results of two land excavations have given different opinions, which are not more than 50 m apart. Interestingly, such an important site does not have scientific dates except one (excavation by Deccan College, Pune, dating Late Medieval Period), which can support archaeological findings.

1.4.3 Offshore Explorations between 1983-1994

The Marine Archaeology Centre of the National Institute of Oceanography, Goa established in 1981, has continuously carried out underwater explorations off Dwarka between 1983 and 1994. The explorations were concentrated in an area of more than 1 square km between water depth of 1 m to 20 m. Several stone structures of well-dressed blocks were noticed in a water depth between 3 and 7 m. They have been described as fortification walls, bastions and pillars, etc. Rao (1990) has dated them to the Protohistoric Period (3500 years BP). A large number of stone anchors were also found from the same vicinity during explorations and they have been compared with stone anchors found at Syria and Cyprus datable to the 12th-14th century BC (Rao, 1987).

The geophysical survey was carried out in 1989 in water depth of 3 to 20 m (Vora *et al.*, 1991). The echograms revealed that the nature of sea bed is generally uneven and rough. The most prominent feature of the region is the existence of a channel observed in the form of symmetrical/asymmetrical cuttings. The channel interestingly shows up prominent V-shape where one flank is steeply sloping. Towards the west, the channel becomes shallower as the sea bed, showing average gradient markedly gentler than the near shore. Bathymetric profiles clearly indicate the change in the nature of bed from near shore to offshore. Towards near shore, the area is rugged and a channel cutting is observed towards the north. Towards offshore the profile shows gradual smoothening of the gradients. The sonographs from the area are typically marked by tonal variations indicating furrows of various sizes and shapes in different directions. Due to this, it is very difficult to isolate any feature, which may indicate man-made activity. Even very thin deposition of sediments as well as sea weeds that are present in plenty on the sea floor can mask such features. However, the sonographs help in demarcating the channel boundary very clearly. The seismic reflection records do not show significant penetration in general except in channel.

The dating of Dwarka has been a major concern for archaeologists and historians during the last several years and contradictory arguments have been put forward on the original location and dating of Dwarka of the Mahabharata period. Okhamandal area (Fig. 2) has been an attraction to early settlers due to its vast potential of marine resources. The earliest settlement dates back to the Harappan Period at Nageshwar, followed by Bet Dwarka, Dwarka, Suvarntirtha and Dhrasanvela, etc. Other villages in the Okhamandal came into existence during the Early and the Late Medieval Period. Therefore, Okhamandal region has been occupied continuously from the Harappan Period till date.

2
Equipment and Methodology

The working environment of underwater archaeology is significantly different than that of land archaeology. It basically deals in sub-tidal regions. Infrastructure facility required for underwater archaeology include a suitable mother boat, a diving boat and other diving systems, which are described below.

Generally, either a fishing trawler or a small passenger boat needs to be hired for the explorations. In case of underwater explorations at Dwarka and Somnath, a fishing trawler named *Gangasagar,* from Bet Dwarka Island, was hired in the seasons 1997-98, 1999-2000 and 2000-2001 whereas in one season (1998-99), a fishing trawler *Vijaya* from Bombay was hired. The vessels were wooden hulled, with good working space especially for keeping diving equipment, LP compressor, SDDE diving equipment, OBM, besides a small cabin for housing VCR, TV, Sony movie camera, underwater metal detector and other cameras with accessories (Pl. 2). Boats were also equipped with marine compass and life-saving appliances, etc.

2.1 DIVING

For underwater archaeological exploration, diving is the primary tool for investigations. The following systems of diving were adopted during the underwater archaeological explorations.

2.1.1 SDDE (Surface Demand Diving Equipment)

This system consists of one low pressure compressor of 25 cfm with filter panel, pressure gauge, demand valve, air hoses and other diving accessories for divers, which facilitate diving for a longer time. For search in poor visibility conditions and for spot work, this system has no other substitute as the diver is constantly in contact with the vessel through the air pipe.

2.1.2 SCUBA (Self-Contained Underwater Breathing Apparatus)

This apparatus consists of a portable cylinder attached with regulator and buoyancy compensator device and other necessary accessories, which can be mounted on the diver's back to enable him to dive freely without any hindrance. SCUBA is simple, yet remarkable, in that it delivers air on demand at a pressure slightly above the ambient pressure. SCUBA system is very useful for underwater surveys especially while operating underwater scooter and underwater camera, drawing, communication system and hand-held metal detector.

2.2 UNDERWATER DOCUMENTATION

Each and every artefact noticed must be recorded before retrieving it. There are various methods adopted to record the *in situ* position of each object, such as videography, photography and drawing.

2.2.1 Videography

For documenting underwater findings, underwater videography on DVC cassette was carried out by using an underwater Subsea Colour Watch, CCD digital video camera with light and an onboard monitor. This system is deployed with an umbilical cable between the camera and TV monitor to view on board the underwater findings as recorded by the camera and also record the same in DVD format. The system also allows underwater communication between the diver and master. The camera has all the control on board with a TV monitor and a wide angle lens with an underwater light, which improves visibility. Other systems for underwater videography were Sony Hi-8 handycam video camera with underwater housing that was deployed for recording with betamax magnetic small cassettes. All the objects were recorded *in situ* before and after cleaning.

2.2.2 Still photography

Another method used for documenting underwater archaeological remains is photography. Underwater Nikon V Camera with 35 mm wide angle lens was used for this purpose. For artificial lighting, a Strobe Gun of 1000 candlelights was used. Depending on the size of the object, a scale of 150 cm, 45 cm or 25 cm length was placed near the object while photographing. Kodak 100, 200 & 400 AS colour film negatives besides a 100 ASA Ectachrome for slides and 400 ASA ORWO film for black & white photography were used.

2.2.3 Drawing

Underwater drawing is a traditional way of documentation. The system consists of an aluminum board (50 x 50 cm) with a graph sheet and a polyester tracing film fixed on the board. Often H or HB pencil is used to draw the figure underwater. It is advisable to draw underwater objects when visibility is better. Each drawing sheet will have date, time, water depth, location and direction with reference to the object.

2.3 GLOBAL POSITIONING SYSTEM

Global Positioning System (GPS) Model Megallan 5000 was used in the survey to obtain the positions. This is a satellite-based radio navigator system designed to provide global position data for navigation. The GPS receives five satellite signals at a time, selects three or four signals with the best reception from among these satellites, and provides the data for position (latitude, longitude and altitude), travel speed and course. The accuracy of GPS varies from 25 to 100 m. This portable system operates on 10-16 V DC power supply.

2.4 UNDERWATER METAL DETECTOR

The metal detector was often deployed from the rubber boat *Gemini*. Different search strategies were worked out considering the sea conditions, swell heights and directions for the survey. A marker buoy was dropped at the location every day and subsequent reference buoys in 50, 100 and 200 m radial distance were dropped in a systematic pattern. The boat was then navigated visually at approximately 5 m spacing grid. It was planned that on a very slow towing boat at the speed of 1 to 2 knots, when a ping is recorded by metal detector a marker buoy would be dropped immediately to pin-point the source and subsequently verification could be carried out by hand-held metal detectors and visual inspection by the divers.

2.5 L.P. COMPRESSOR FOR AIRLIFT OPERATION

The ELGI 125 cfm air compressor consists of two main components, namely, compressor and engine. It requires 12 volt battery to start the engine and runs on diesel. It can also be operated by 3-phases eletrical power supply. The electric motor rotates the compressor so as to get sufficient output of air. The compressor part is oil lubricated with two cylinders, which are capable of discharging air at 125 feet^3/minute. The maximum rate at which the compressor lets air is 7 kg/cm^2. This compressor can be used for airlifting from 3 to 50 m water depth.

METHODOLOGY

Marine archaeological explorations are generally initiated with coastal explorations followed by offshore explorations. The details of the explorations at Dwarka are described below.

2.6 ONSHORE EXPLORATIONS

Onshore explorations were carried out first to understand and select the important archaeological locations for underwater explorations. While carrying out coastal archaeological surveys, features such as cliff sections, etc. containing evidence of human habitation were examined which can be related with the marine archaeological studies. These explorations were carried out through observations along the coast and important archaeological findings were documented with the help of photography and drawing. Positions of these locations were also recorded by GPS.

2.7 OFFSHORE EXPLORATIONS

2.7.1 Underwater visual survey

The diving operations were initiated at Somnath in 6 m water depth towards north of Somnath temple and gradually moved towards south. The operations at Dwarka were initiated off Samudranarayan temple and gradually extended along the coast as well as in deeper waters. Each diving covered the area within a circle of about 50 m radius in a circular search pattern. Diving operations were also carried out at several places in shallow waters from 4 m to 8 m and also at eight locations in deeper waters (10 m to 23 m). A buoy was tied to each location wherever archaeological objects such as structures and anchors were noticed. The objects were inter-linked underwater by a connecting rope. The distance between two objects and orientation were taken by underwater compass. Underwater findings were recorded by a still camera and each object was drawn as per scale separately.

2.7.2 Airlifting operations

Airlift is a system which felicitates excavations of sea floor to know the stratigraphy and to reveal the buried antiquities. Powerful suction was used for the excavation, which consisted of a 10 cm diameter pipe with a direct connection of air supply by pressure compressor through a 6″ diameter pipe. A 12″ long chamber was welded at 9″ above the bottom of the main pipe. The main pipe had 1/8″ diameter perforation within the chamber. When air at about 90 CFM or more is supplied through the chamber, the perforation causes split in the air, forcing it to form air bubbles, which move upward within the main pipe. Buoyancy is created at the sea but due to the air-water mixture helping in the lifting up the sediments and particles from the seabed through the main pipe. This water and sediment mixture is collected in a container onboard the survey vessel while passing through a sieve, which connects sediments and other solid materials and the water is allowed to drain out into the sea again. Antiquities may be found in the sediments and solids retained on the sieve.

3
Underwater Explorations

Marine archaeological explorations at Dwarka comprise onshore, intertidal zone and underwater investigations. Onshore exploration has been carried out between Rupen Bandar on the north and Okhamadi on the south of Dwarka. Intertidal zone explorations were undertaken between the lighthouse in the north and Okhamadi in the south during low tide. Offshore explorations were restricted to Dwarka waters only. A detailed description of the findings is given below.

3.1 ONSHORE EXPLORATIONS

In the course of onshore exploration a huge sand dune was noticed on the mouth of the river Okhamadi and it is surmised that the river was navigable in the past. However, no archaeological objects were noticed around this area. Two temples of the Solanki period were noticed about 20 km south of Dwarka. These are locally known as Shamb Lakshamana. The temples were constructed out of calcareous stone blocks on a high platform and beautiful sculptures were engraved on the walls of the temples.

A single-holed stone object spheroidal in shape was noticed on the southern side of Dwarka coast and that is lying well above the high water line. The object has an across circular hole in the middle with chisel marks on it. It is made of conglomerate rock with a reddish appearance. Seven stone anchors were noticed at a depth of 0.5 m to 1 m during low tide. Three of them are triangular and others are grapnel type. All the anchors are made out of limestone. One of the grapnel anchors is the heaviest among stone anchors so far recorded in Dwarka area. It has two rectangular holes: one on the lower side and a circular hole on the upper side. Another anchor found at a depth of 0.5 m is a heavy one with two square holes on the lower side and an upper circular hole. The triangular anchor was lying in 0.5 m water depth. It was covered with seaweeds. The anchor is intact with two rectangular holes on the lower side and an upper circular hole. Other two triangular anchors are broken on the lower side.

During the field season, 2000-2001, four circular stone structures were noticed in the intertidal zone of Dwarka during low tide (Pls. 4 & 5). These are similar to those found offshore

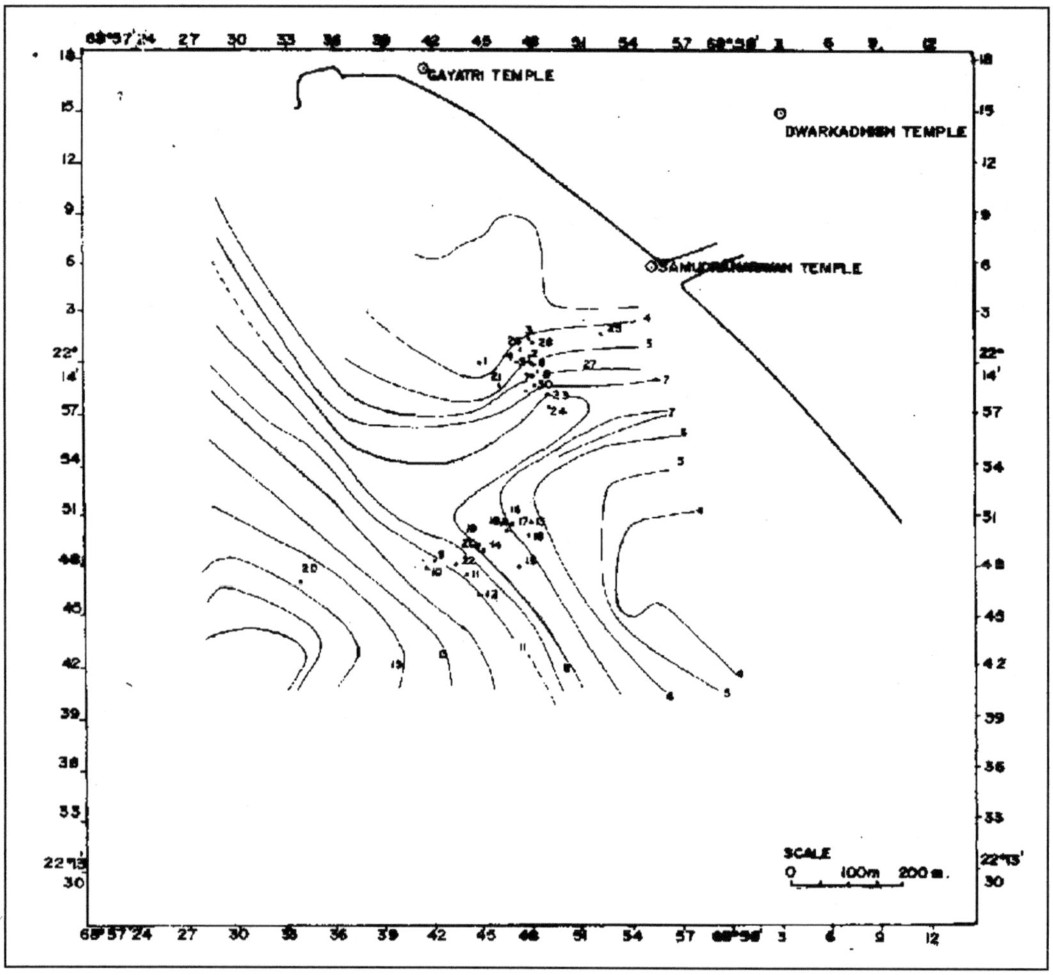

Fig. 3. Underwater contour map off Dwarka based on echo-sounder profile.

of Dwarka. They are lying adjacent to the mouth of the Gomati creek. A large number of rectangular blocks were also noticed. All these circular structures are in one line. The diameter of one structure is about 2 m. A large area gets exposed during low tide in Dwarka region (Pl. 6).

Another important site at Rupen Bandar, 2 km north of Dwarka, has the remains of a jetty. The collapsed remains of a jetty get exposed during low tide (Pls. 7 & 8). Three rows of pillars have been noticed right from the high water line to 100 m offshore (Pl. 9). The structure is made out of dressed limestone blocks. A wooden pole has been noticed in the centre of each pillar of the jetty. According to local information, this jetty was used for loading of cement and also used by the fishing trawlers. No stone anchor was reported from this harbour.

3.2 OFFSHORE EXPLORATIONS

Offshore exploration included geophysical survey and underwater visual survey. The findings are given below.

3.2.1 Echo-sounder survey

Echo-sounding was carried out in the area mainly to obtain first-hand information about seabed characteristics, especially the presence of geomorphic features, if any, rather than bathymetric.

Though GPS was used to obtain the positions, the lines were actually transacted not by pre-plots but by the shore objects and the marine compass was installed on dinghy. The tidal correction was not carried out on the data collected; the same has not been used for bathymetric purpose. Main features observed qualitatively are as follows:

1. A relatively deeper depression was present south of the lighthouse.
2. Presence of a 2 m scarp was observed on the west of the lighthouse on 2-3 courses of entire line.
3. At times decrease in seabed gradient was observed at 13-14 m water depth.
4. Beyond 16 m water depth the seabed was gently steep without much unevenness.

Submerged channel of the Gomati, reported earlier (Vora *et al.*, 1991), was located again at some distance away from the shore (Fig. 3). However, the same was not present near the shore, as it was probably filled with sediments. It was possible to identify the nature of seabed and channel characteristics broadly on the basis of echo-sounding image tones.

3.2.2 Underwater visual explorations

Extensive diving operations were undertaken between 1998 and 2001 in various places and depths ranging from intertidal zone to 25 m waters off Dwarka (Fig. 4). The scanned area

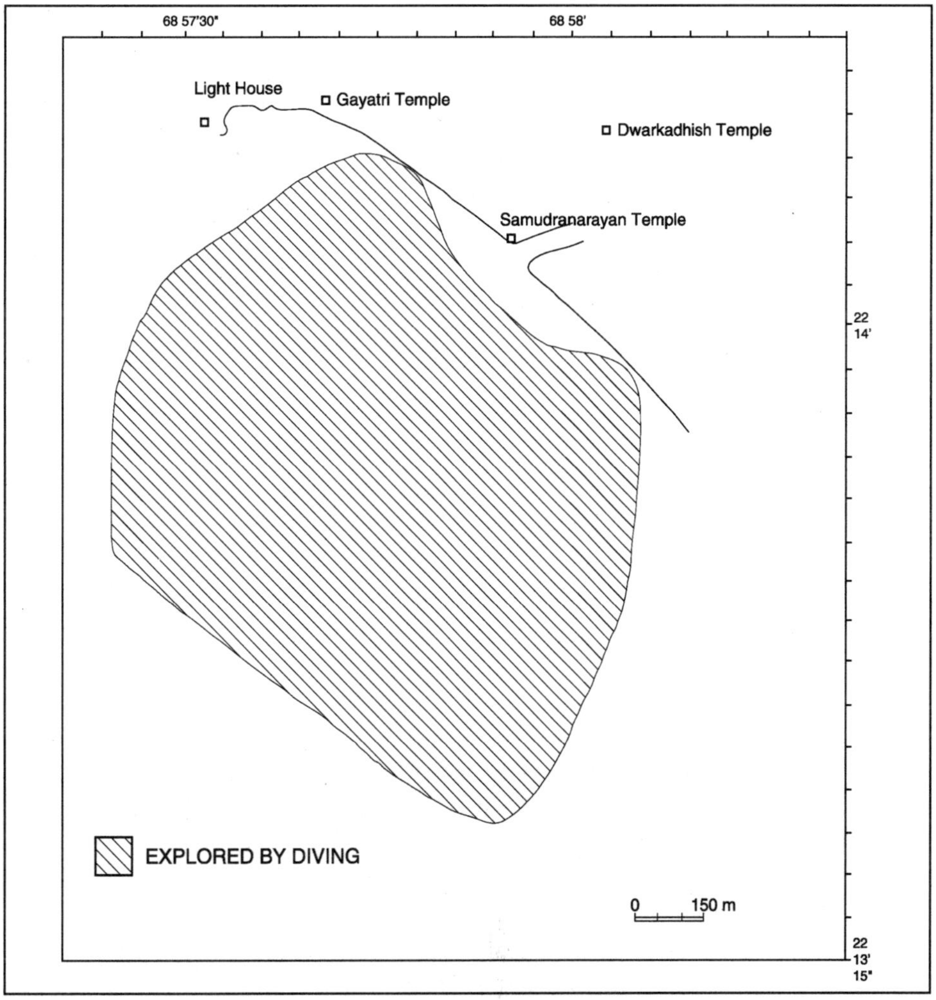

Fig. 4. Underwater diving area off Dwarka.

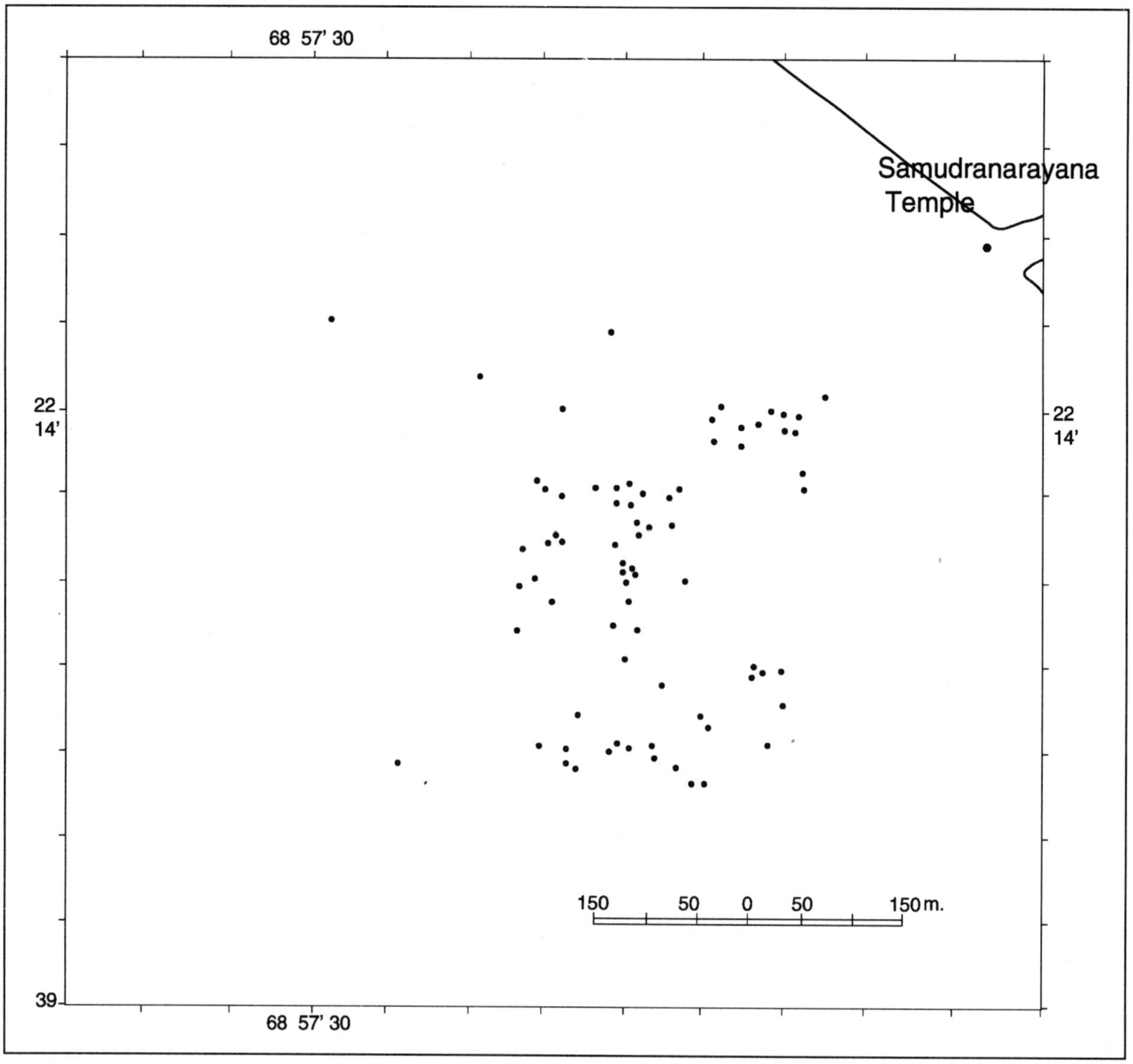

Fig. 5. Location of archaeological artefacts off Dwarka.

approximately measures 1 km x 1 km. On the basis of archaeological findings (Fig. 5), the entire area has been divided into four zones. A brief description of the findings is given below.

3.2.2.1 Zone 1

The site is lying opposite the mouth of the Gomati creek and southern side of the transit line of the Dwarkadhish and the Samudranarayan temples (Pl. 3). The location is about 200 m offshore from Samudranarayan temple in the water depth ranging from 3 m to 5 m. The topography of seabed consists of natural beach rock formation, which is covered with densely grown vegetation/ seaweeds. Thick deposition of sand was also noticed in the small channels. A large number of scattered stone blocks were noticed (Fig. 6) and many of them are found partially buried in sediments. Several semicircular structures (Pls. 10-11) were noticed and a few of them are fully scattered. The blocks of semicircular structures have L-shaped cut (Pl. 12) besides provision for dowels and a few of them have cementing substances, which have bounded them even to this

Pl. 1. Jagatmandir, Dwarkadhish.

Pl. 2. Boat used for offshore explorations.

Pl. 4. Circular stone structure exposed during low tide in intertidal zone of Dwarka.

Pl. 3. Location of underwater site with reference to Dwarkadhish and Samudranarayana temples.

Pl. 5. Damaged circular stone structure exposed during low tide in intertidal zone of Dwarka.

Pl. 6. A vast area is exposed during low tide in Dwarka.

Pl. 7. A 120-year-old jetty is exposed during low tide at Rupen Bandar, about 2 km north of Dwarka.

Pl. 8. Remains of a circular structure, similar to that noticed at Dwarka, is exposed during low tide at Rupen Bandar, about 2 km north of Dwarka.

Pl. 9. Three rows of wooden pole, the remains of a 120-year-old jetty, were exposed during low tide at Rupen Bandar, about 2 km north of Dwarka.

Pl. 10. Remains of a circular structure lying in 5 m water depth off Dwarka.

Pl. 11. Remains of a circular structure lying tilted in 5 m water depth off Dwarka.

Pl. 12. Block with semicircular structure with L-shaped cut and curve noticed off Dwarka.

Pl. 13. Partially buried structure lying in 5 m water depth off Dwarka.

Pl. 14. Stone structures covered with dense vegetation off Dwarka.

Pl. 15. An alignment of stone blocks noticed in 5 m water depth off Dwarka.

Pl. 16. A dressed stone block with a projection lying off Dwarka.

Pl. 17. Scattered stone blocks in 5 m water depth off Dwarka.

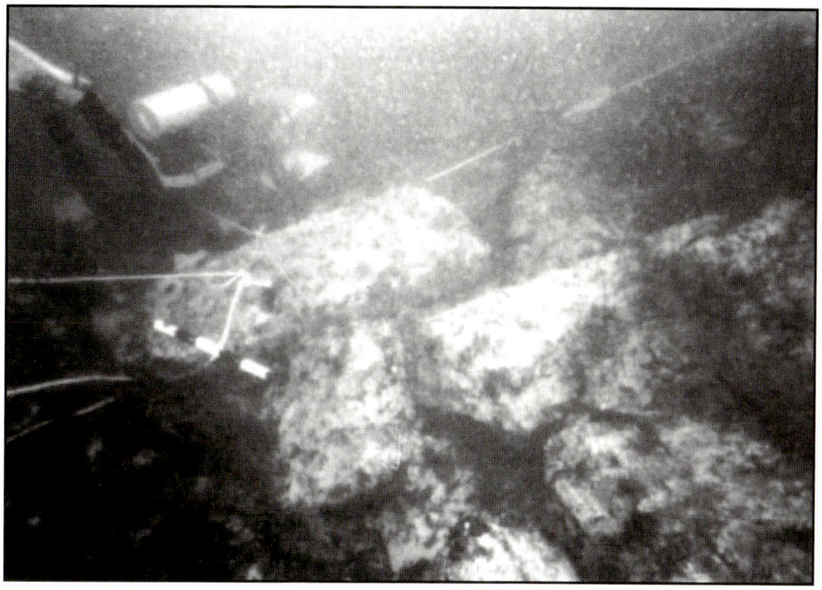

Pl. 18. A dressed stone block with a projection found off Dwarka.

Pl. 19. Stone block with a Gujarati script noticed off Dwarka.

Pl. 20. Submerged cliff at 8-10 m water depth off Dwarka.

Pl. 21. The biggest composite stone anchor found off Dwarka.

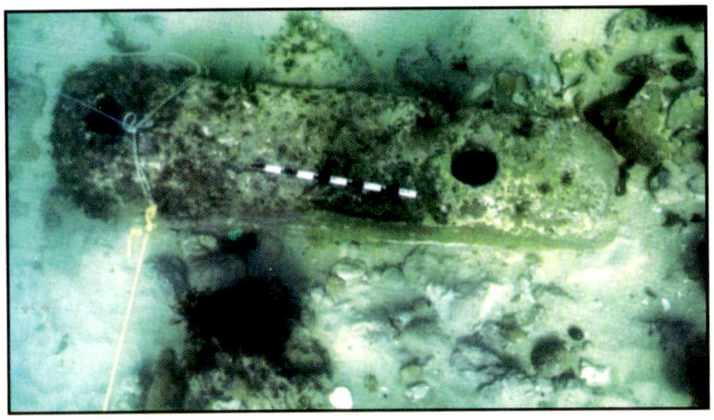

Pl. 22. Indo-Arab type stone anchor lying off Dwarka.

Pl. 23. Ring stone type stone anchor lying off Dwarka.

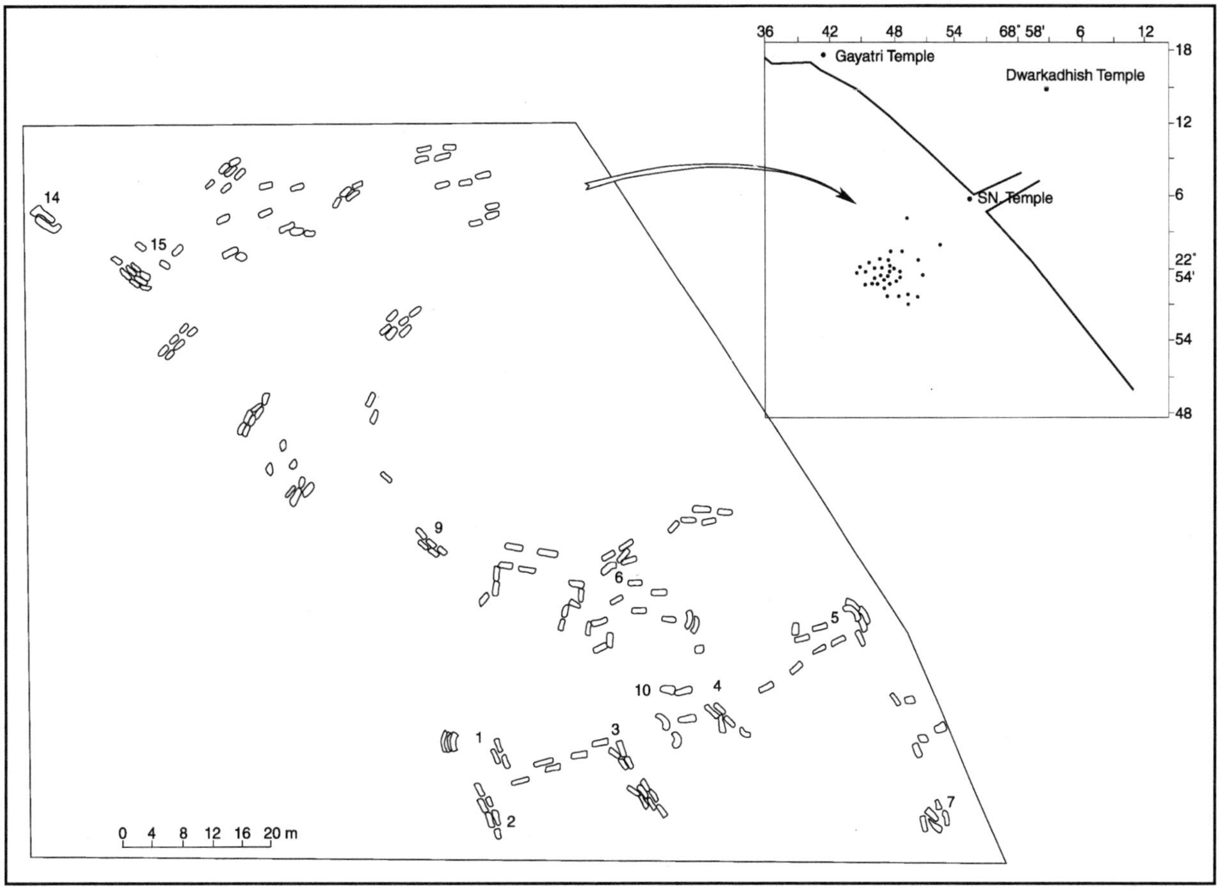

Fig. 6. Location of structural remains in shallow water between 3 to 5 metre water depth, Zone 1.

day. These structures have 2 to 3 courses with a 60 to 80 cm height. The average size of one block of semicircular structure is 95 x 55 x 25 cm.

Besides semicircular blocks, a large number of rectangular blocks have been observed in this area (Pls. 15-18). They are scattered in a large area and do not form any kind of regular form / shape at present. However, the presence of semicircular structures near these blocks indicate that they are the part of the same structures. The rectangular blocks have various size and important ones are 120 x 60 x 16 cm, 115 x 50 x 17 cm, 110 x 50 x 18 cm, 85 x 48 x 16 cm, and 45 x 30 x 15 cm (Fig. 7). These blocks are lying on the rocky seabed and a few of them are buried in sand (Pl. 13). There is a channel on the northern side in which a few structures were noticed. The exposed portion of these blocks is covered with thickly grown seaweeds (Pl. 14) and a pinkish layer of marine growth. The important findings may include a rectangular stone block bearing Gujarati script (Pl. 19 and Fig. 8) which may not be very old.

There were 13 stone anchors noticed from Zone 1 close to the scattered blocks. Among them 3 anchors are of triangular type, 7 of grapnel type and 3 of ring stone anchor type. These are of various sizes and all of them are made of limestone. Triangular anchors have an upper circular hole and the lower two holes are square having even thickness. Similarly, grapnel anchors have a circular upper hole and the lower two holes on either side are rectangular. Three ring stones are significant in this area as these were earlier reported beyond 8 m depth in Somnath and Dwarka waters. The vegetation growth has been noticed over the anchors.

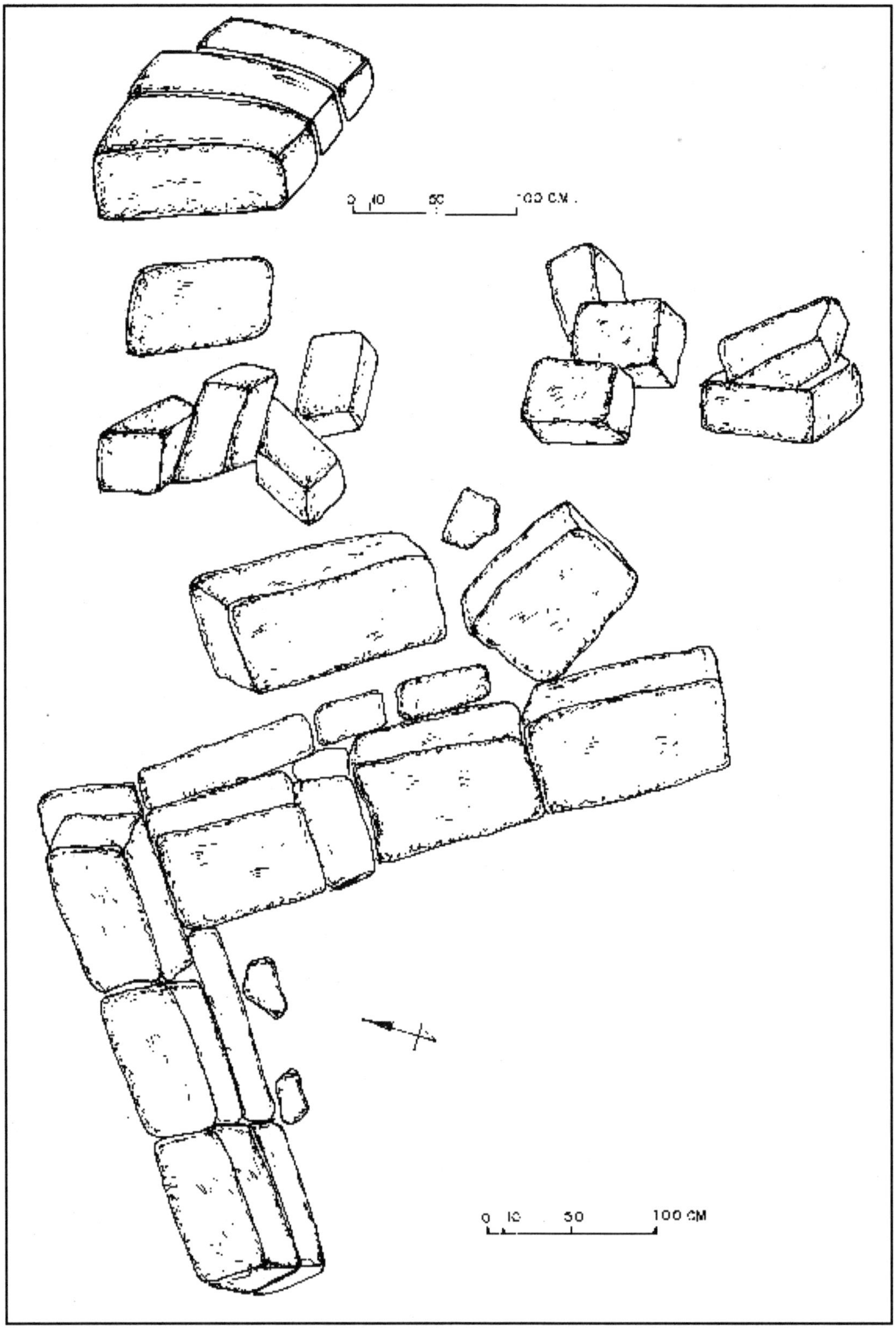

Fig. 7. Underwater plotting of stone structures noticed off Dwarka.

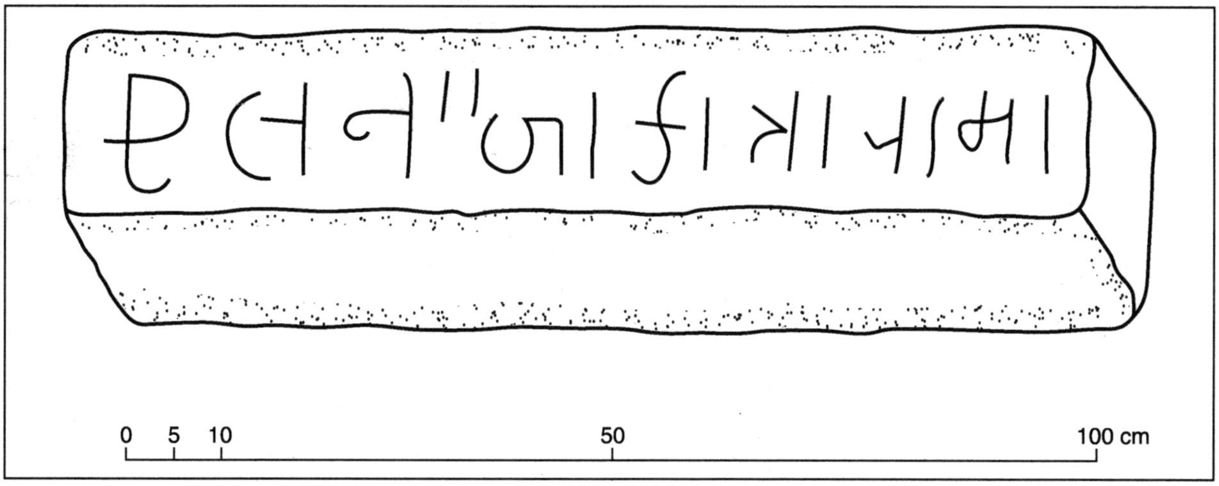

Fig. 8. A stone block with graffiti.

3.2.2.2 Zone 2

The area is located westward of Zone 1 on the transit line of the Dwarkadhish and the Samudranarayan temples. The water depth ranges from 5 to 8 m. The topography of seabed in this area consists of the exposure of rock bed and a few sandy patches. The exposed rock is covered with thick vegetation.

Stone anchors are scattered over a large area in this zone. In all, 34 stone anchors of different types and sizes were recorded (Fig. 9) which included 7 composite, 5 grapnel and 12 ring stone

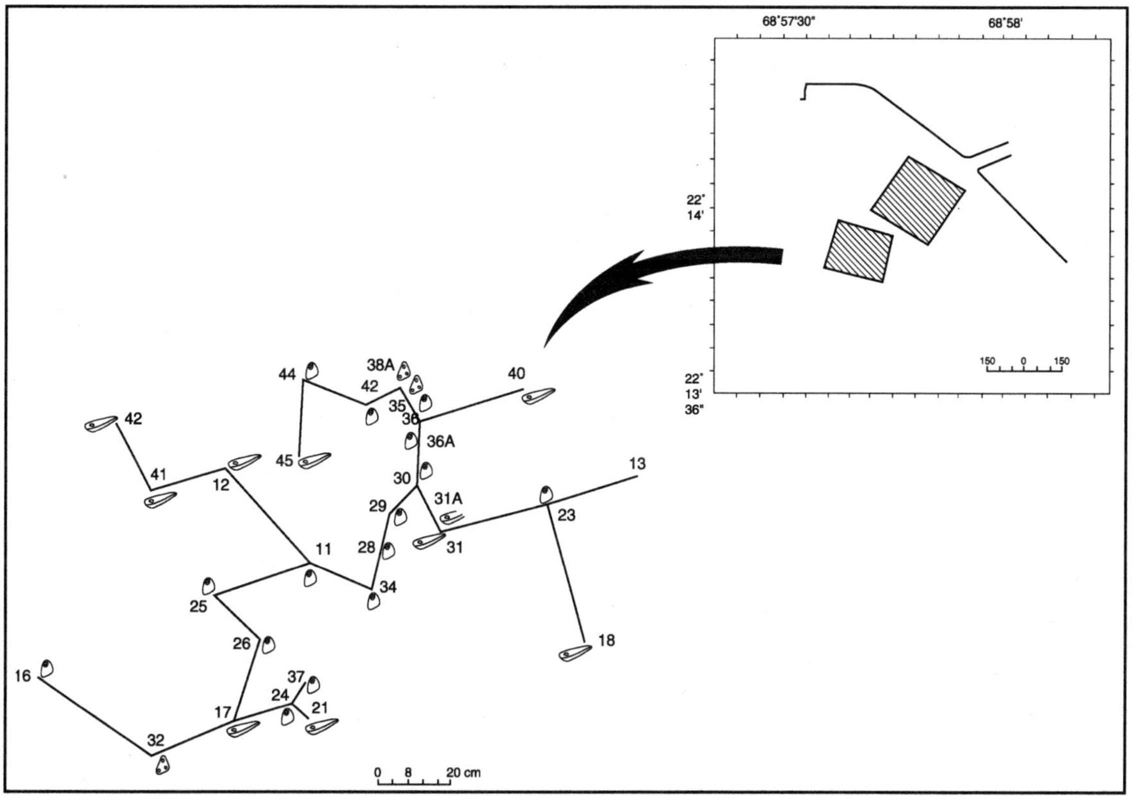

Fig. 9. Underwater plotting of stone anchors between 5 and 7 m water depth off Dwarka.

type. A few anchors are partially buried in sediment; however, a majority of anchors are lying exposed over the rocky seabed. Each exposed anchor has a thin layer of pinkish seaweed growth. One of the biggest triangular stone anchors (Pl. 21) has been reported from this area and similar example have not been reported from anywhere in India. Similarly, two grapnel anchors of 2.3 m length have been noticed from this zone (Pl. 22), which are perhaps the heaviest anchors found in Dwarka waters. All the triangular and 13 grapnel types of anchors are made of limestone while 2 grapnel anchors are made of basalt.

Twelve ring stone anchors of different shape and sizes were noticed in this area. The common shape of ring stone is consisted of a circular flat bottom which tapers on the upper side with a circular excel hole in the centre (Pl. 23), while the other type includes both ends flat with a rough and porous surface. The anchors are made of limestone.

3.2.2.3 Zone 3

The area falls in the southwest direction of Zone 2 and southern side of the transit line of the Dwarkadhish and the Samudranarayan temples. The seabed here consists of rocks and a ledge in this area and in the southern side, a deep channel with fine sand. The airlift operation in the channel revealed deposition of sand, clay, rubbles and rock bed in sequence. In this area marine growth was negligible, while anchors were covered with a thin layer of greyish marine growth. A large number of anchors were reported from this zone and they are lying exposed on a rocky bed, trapped in-between rock and sometimes partially buried in sand. The total 28 stone anchors of different types and sizes are reported which include 20 anchors of grapnel type and 8 of triangular type. All of them are made of limestone except one grapnel anchor which is made of basalt rock. An interesting aspect of a triangular anchor is the presence of a rectangular hole on the upper side, which is unique and not reported so far; however, its original upper circular hole is broken. The anchor is made of fine grained limestone with even thickness. Another triangular anchor has an additional circular hole in the middle, which is also a new phenomena and not reported from anywhere esle. However, recently similar types of anchors have been reported from Bet Dwarka waters (Gaur *et al.*, 2005). Four ring stone anchors were also reported from this zone. Two were lying exposed on the seabed while the other two were partially buried in sediment in the channel and were exposed by using airlift. A ring stone is made of basalt while the other three are of limestone.

3.2.2.4 Zone 4

The area is situated on the northern side of the Dwarkadhish and the Samudranarayan temple transit line. Water depth ranges from 3 to 8 m. The seabed is comprised of rock formation with channels where sand has been accumulated and a few anchors have been trapped in-between the rock boulders. Six stone anchors of different type have been reported from this zone. All of them are made of limestone. Among them two are triangular type and one of them is broken. Three grapnel type anchors are intact; however, one of them have been integrated with rock bed and the other one is lying in channel which was partially buried in sand. A ring stone anchor was also noticed on the northern side towards the lighthouse. This is small in size compared to the ring stones of Zone 2 and 3. The area is overgrown with seaweeds and other marine growth.

3.2.3 Diving at deeper places

As stated earlier, diving was undertaken between 16 and 25 m water depths to ascertain the presence of underwater structures. A brief description of the same is given here.

Diving at 20-25 m parallel to the shore yielded a strip of rocky formation over a width of 10 to 15 m. At a few places it is raised up to 1 m and at others it is buried below 5 to 10 cm thick sand. Towards the shore from this point, a few rocky patches with white sand were noticed.

The southeast of Zone 1 area, in 2-8 m water depth, revealed rock formation with heavy growth of seaweeds. Opposite to this area, towards the shore, a few stone anchors were also noticed in intertidal zone.

The explorations off Dwarka brought to light a large number of artefacts such as anchors related to the maritime activities and navigation. The observations and views over the underwater findings have been discussed below.

3.4 DISCUSSIONS

Extensive explorations in a large area of Dwarka brought to light a number of structures and artefacts between intertidal zone and 16 m water depth. These include semicircular structures, rectangular and square blocks, different types of anchors and single-holed stone objects. A brief discussion on the importance of the findings is given below.

3.4.1 Structures

Submerged stone structures have been noticed just opposite the mouth of the Gomati creek between intertidal zone and 6 m water depths. The structures are lying in a vast area (225 x 275 m) and are fully scattered. They include semicircular structures and dressed blocks of different sizes.

3.4.2 Semicircular and Circular structures

These structures were noticed at different places between water depths of 3 and 6 metres. They were lying scattered. At times, the structures are intact but generally found lying tilted. On an average, they have 2 to 3 courses. The blocks used for these structures are similar to those found in the intertidal zones of Dwarka. However, the circular type structures from intertidal zones suggest that the offshore structures may have been circular in shape earlier, and later they collapsed and the remnant form is semicircular one. The study of the structure of intertidal zone suggests that dressed blocks were used on the outer side and the centre was filled with irregular blocks with a binding material. The position of the structures in intertidal zone indicate their linear alignment; however, offshore structures are disturbed and it is difficult to suggest any alignment.

It has been suggested that these structures could be the bastions for some fortified structure of a habitational site (Rao, 1990, 1999). However, it is important to note that the basic requirement for a habitational site in archaeological terms, such as pottery, coins, seal or any other day-to-day use items, has not been recovered so far from the offshore structures even after extensive airlifting in the vicinity. Interestingly, a large number of similar kinds of structures have been observed near Rupen Bandar, a site 3 km north of Dwarka. They are part of a jetty, which has now collapsed and lying submerged in the sea and which gets exposed during low tide. There are no stone anchors from Rupen Bandar whereas stone anchors along with stone structures in Dwarka waters suggest that the structures of Dwarka are older than those reported from Rupen Bandar, which may be dated to the early 19th century AD. In this circumstance, the structures reported from Dwarka might have been the remains of an ancient jetty.

The study of protohistoric settlements in India suggests that, in the past bastions were generally constructed only in very important cities and not in each and every town or village.

One of the most important site of the Harappan Period, Lothal does not have a single bastion on its defence or fortification wall (Rao, 1979). Similarly at Nageshwar, a mature Harappan site, does not have any indication of fortification or bastions (Hegde *et al.*, 1990). On the other hand, Dholavira (Bisht, 1990), Surkotada (Joshi, 1990), Kuntasi (Dhavalikar *et al.*, 1996), Bagasra (Sonawane *et al.*, 2004) and a number of other important Harappan sites have square or rectangular bastions. Interestingly, in Dwarka the so-called bastions are circular while in Harappan settlements they have rectangular and square in shope. In comparison Dwarka structures do not belong to any fortification dating back to the Protohistoric Period. The discovery of rectangular stone blocks with Gujarati script from offshore Dwarka suggests that some of these structures may be of the Late Medieval Period.

3.4.3 Dressed stone structures

Along with circular and semicircular structures, a large number of dressed stone blocks are observed off Dwarka. They are found scattered in a very large area. They are of various sizes but thickness is same. These structures are identified as remnant of a fortification wall (Rao, 1988, 1990 and 1999). Similar kind of dressed blocks have been noticed in Rupen Bandar as well, which form the structure between two pillars. As mentioned earlier, so far there is no evidence of any habitation from Dwarka. As suggested above, if semicircular structures were part of pillars of a jetty, then these dressed blocks could be part of the super structure of the same.

Dwarka being situated on the open sea, is prone to storms and coastal erosion. The examples from other coastal sites of ancient period suggest that often ancient ports were situated in sheltered places such as on the island, bank of creeks and backwaters.

3.4.4 Harbour or Habitation?

As stated earlier, the structures are of fine dressed blocks and some strong binding material is present. These structures were identified as palace, fortification walls, bastions and gateways (Rao, 1999). However, evidences of pottery and other archaeological material, which are generally reported from archaeological sites, are lacking. For instance, a large number of pottery have been recovered from 5 to 8 m waters off Poompuhar (Gaur and Sundaresh, 1998) in the Bay of Bengal which is known for a turbulent weather condition. Many researchers have criticized the dating of structures off Dwarka (Kenny, 1993; Tripathi, 1996) but not the function of these structures.

The circular structures in the intertidal area suggest that they are the base of pillars while bastions are attached with the fortification wall and used as support for the main wall. Therefore, these structures cannot be considered as the bastion of a fortification. Similarly, while comparing with other Protohistoric sites in Saurashtra, such fine dressed stone blocks (like of Dwarka) have not been noticed elsewhere. Therefore, in the absence of habitational evidences, and nature of structures, we hypothesized that the stone structures found off Dwarka are part of a jetty. The antiquity of these remains is discussed below.

3.4.5 Probable date

These structures have been variously dated from the Protohistoric Period (Rao, 1990, 1999) to the Historical Period (Tripathi, 1996) by juxtaposing findings off Dwarka with those from Bet Dwarka, a site 30 km away, without giving any conclusive argument. Bet Dwarka has well-defined evidence of the Protohistoric settlement. On comparison of structures from both sites, one finds that the structures of Dwarka comprise very fine dressed stone blocks and the binding material seems to be lime mortar while the stone blocks used for Bet Dwarka structures on land and intertidal zone are not dressed properly and the binding material used is clay mortar. Further, bastion has not been observed in Bet Dwarka Island. Therefore, the evidence from Bet Dwarka Island cannot be compared with Dwarka structures and both have to be considered on merit independently.

It has been observed that due to one or the other reason, the beach sand at Dwarka consolidate very rapidly and the structures of the early 20th century AD are buried under well-consolidated beach sand in the intertidal zone and structures of 3500 years should be buried further below instead of on the surface which are lying exposed over the rocky and sandy bottom. Of course, sedimentation regime changes with the depth and it is possible that stronger currents may prevent deposition of sediments. In view of this, there is possibility that the underwater structures reported off Dwarka may not be very old and form the part of a jetty for handling cargoes and anchoring ships and boats in the recent past.

The findings and dates of Dwarka have evoked keen curiosity in the scientific community. At times methodology has been severely criticized as L.B. Kenny (1993:99) opines, "unless archaeology, an auxiliary science of history, is used scientifically, along with literary sources, the excavations of Dwarka would continue to be pseudo-scientific as they appear to be today. History is interpretation based on human reasoning, not on emotion." In light of such strong criticism regarding the validity of date for structures found in submerged Dwarka, there is an urgent need for continuing the excavation on scientific lines to arrive at a logical solution.

3.4.6 Submerged Cliff

An interesting finding is the submerged rocky cliff (ridge) in water depth of 7-10 m (Pl. 20). The cliff is oriented in the north-south direction parallel to shoreline. The highest point of the cliff section is about 1 m from the seabed. Rao (1999) thinks that the harbour was situated further seaward from where a submerged cliff has been reported. The water depth at this place varies between 7 and 10 m depending on the tide and the maximum height of the cliff does not exceed 1 m. In this situation two questions need to be answered: (i) was the sea level lower by 10 m during the mid-2nd millennium BC? and (ii) does 1 m high cliff serve as harbour in open sea coast?

The studies of sea level fluctuation across the globe suggest that sea level was lower by 100 m or more between 17 kyr and 12 kyr and thereafter started rising rapidly and reached to the present level or even little higher during the mid-Holocene (Fairbridge, 1960 and Fairbank, 1989). During this period, major civilizations of the world came into existence such as the Mesopotamian, Egyptian and the Indus Civilization. After attaining the present sea level around 6000 years BP, minor sea level fluctuations (not more than 2 to 4 m) took place till the Medieval Period. Similarly, the studies of sea level fluctuation of Gujarat also indicate that sea level was higher during the period of the Indus Civilization (Merh, 1992; Gaur and Vora, 1999). Accordingly, it is possible that some of the sites located in hinterland must have served as tidal ports. The coastal excavation at Bet Dwarka has revealed the evidence of lower sea level by 2 metres around the Christian era (Gaur *et al.*, 2007). Such evidence from close-by archaeological sites along with other geological parameter, do not support a 10 m lower sea level during the mid-2nd millennium BC; therefore, the cliff at water depth of 7-10 m may not be associated with the harbour of the 2nd millennium BC.

The second question is the height of the cliff is 1 metre and the tidal variation in this area is more than 3 m. Under these circumstances, most of the time cliff will be exposed and therefore cannot serve as a natural harbour. Underwater cliffs are generally the remnant of ancient shorelines – those that have submerged in due course of time.

4
Stone Anchors

The anchor is one of the main implements used in navigation. It may be made of stone, lead or iron. Under water, the most preserved remains of ancient maritime activity could be the stone anchors of different types, as every boat requires an anchor to halt in mid-sea. Initially, the idea of using stone as material for anchors would have come easily due to easy availability and possibility of working on them as required. The convenience of dispensing with them as per circumstances was an added benefit. The large number of stone anchors found from the area confirmed the fact that hectic maritime activities were going on even before the use of lead or iron anchor came into practice. The types and sizes of anchors indicate the kind of boat that visited the port (Frost, 1991), and with the help of these anchors it is possible to trace the authenticity of literary references about the trade routes and the identification of a particular port. Initially, the anchor was a simple stone tied onto a rope; it probably began to develop with experience in navigation. Later on, the shape and size of anchors also changed in line with the requirement to anchor ships in the varying topography of the seabed (Gaur *et al.*, 2001).

Archaeological evidence has revealed that stone anchors are as old as the civilization datable to the third millennium BC (Frost, 1970; Rao, 1979; Dhavalikar *et al.*, 1996; Gaur and Sundaresh, 2005). Stone anchors have different shapes and sizes and have evolved over a long period of time.

4.1 ANCHORS IN ANCIENT LITERATURE

The Sanskrit word for anchor is *Nangara*. The *Milindapanho* (Max Muller, 1894), a Pali text of the early centuries of the Christian era, has interesting conversation on anchors between King Milind and the Buddhist monk Nagasena. As the king asked, "Venerable Nagasena, those two qualities of the anchor which you say he ought to take, which are they?" Nagasena described the qualities as,

> Just, O king, as the anchor, even in the mighty sea, in the expanse of waters agitated by the crowding of ever varying waves will fasten the ship, and keep it still, not letting the sea take it in one direction or another; just so, O king, should the strenuous Bhikshu, earnest in effort, keep his mind steadfast in

Pl. 24. Composite or triangular type stone anchors No. 1 to 10 from Dwarka waters.

DT-11

DT-12

DT-13

DT-14

DT-15

DT-16

DT-17

Pl. 25. Composite or triangular type stone anchors No. 11 to 17 from Dwarka waters.

DT-18

DT-19

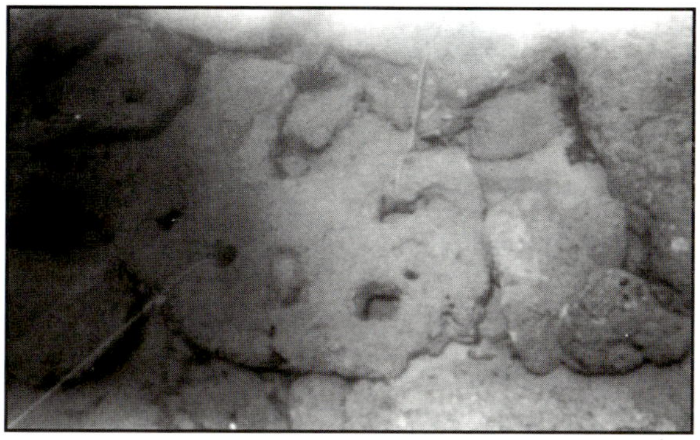

DT-20-22

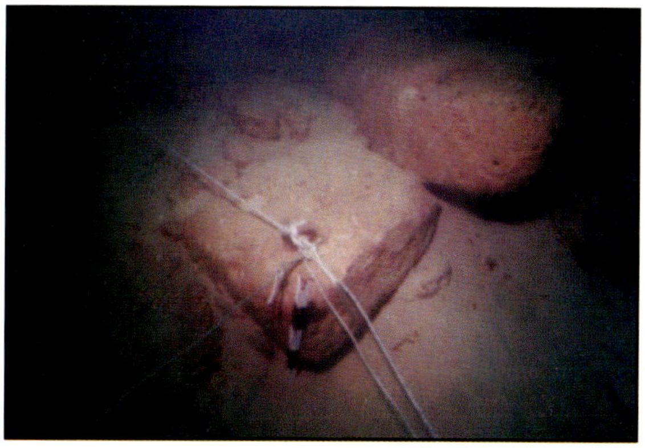

DT-21

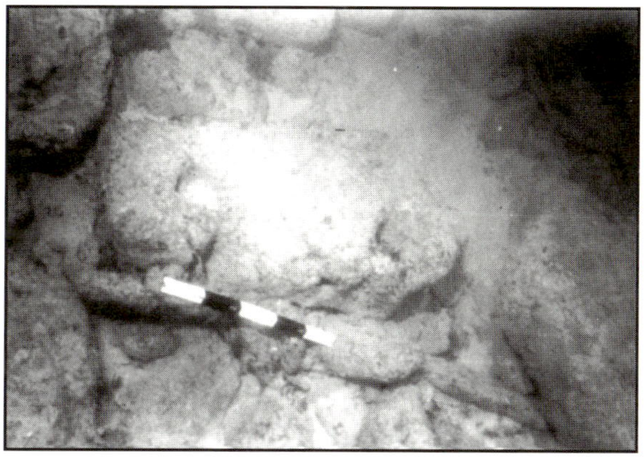

DT-23

DT-24

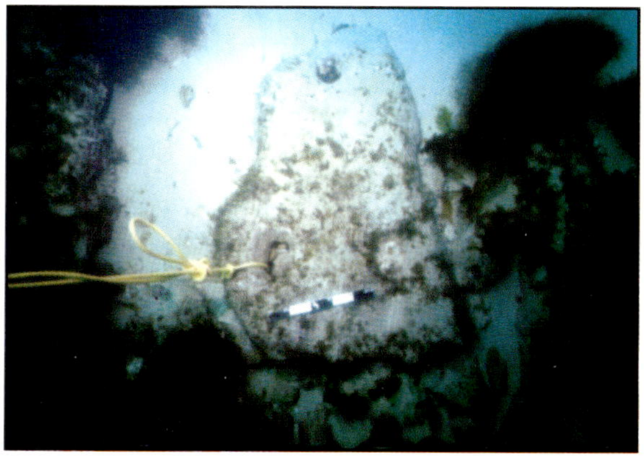

DT-25

DT-27

Pl. 26. Composite or triangular type stone anchors No. 18 to 27 from Dwarka waters.

DT-26

DT-32

DT-34

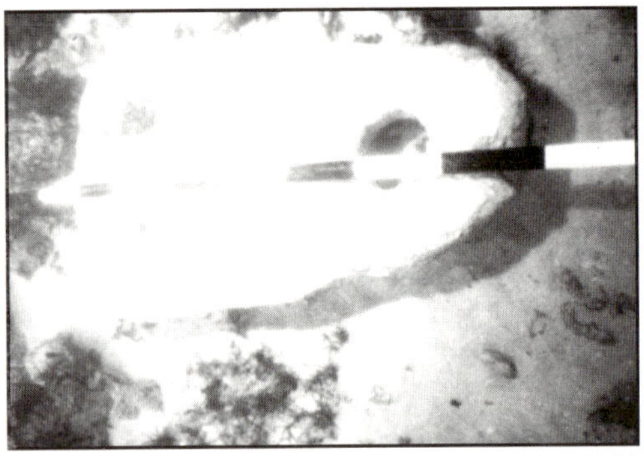

DT-28

DT-29

DT-31

DT-35

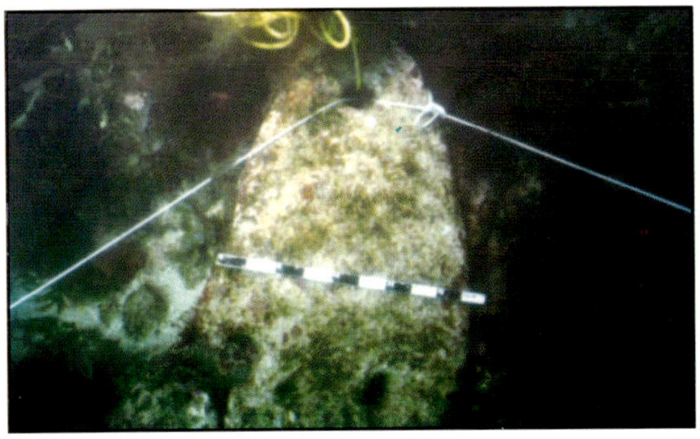

DT-33

DT-36

Pl. 27. Composite or triangular type stone anchors No. 28 to 36 from Dwarka waters.

DG-1

DG-2

DG-3

DG-4

DG-5

DG-6

DG-7

DG-8

DG-9

DG-10

Pl. 28. Indo-Arab type of stone anchors No. 1 to 10 from Dwarka waters.

DG-13

DG-11

DG-12

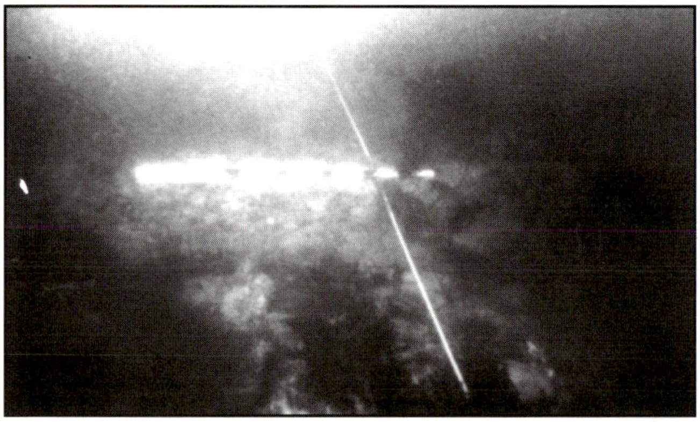

DG-14

DG-15

DG-16

DG-17

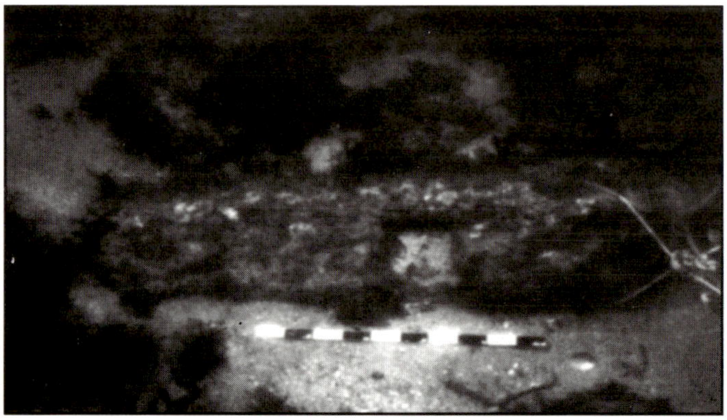

DG-18

Pl. 29. Indo-Arab type of stone anchors No. 11 to 18 from Dwarka waters.

DG-19

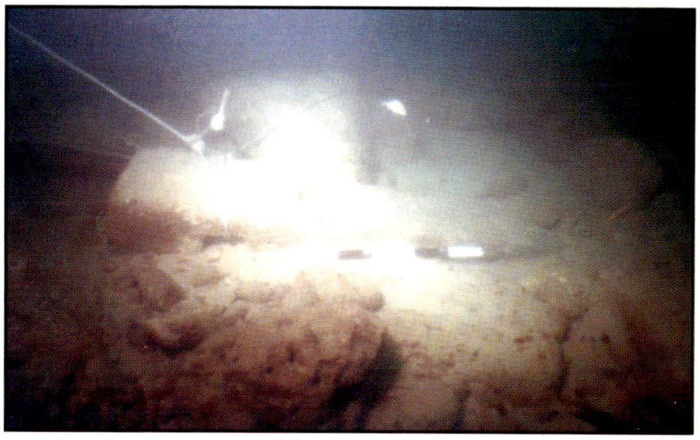

DG-20

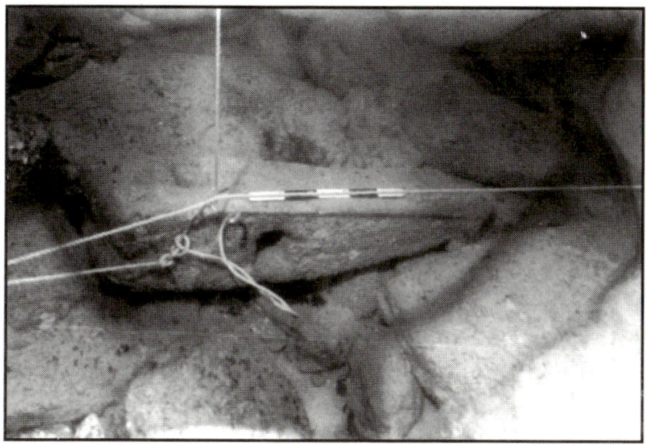

DG-21

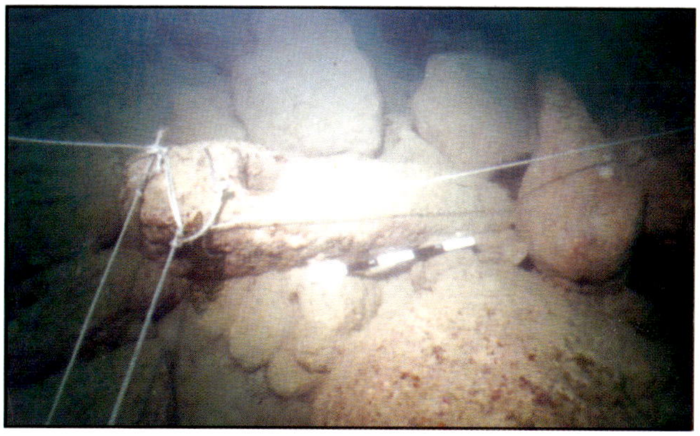

DG-22

DG-23

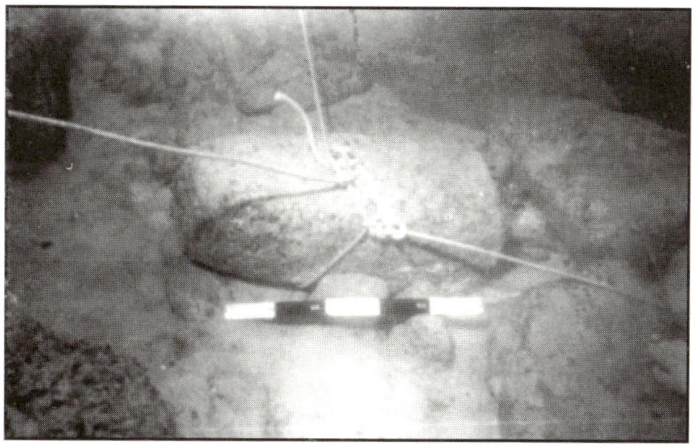

DG-24

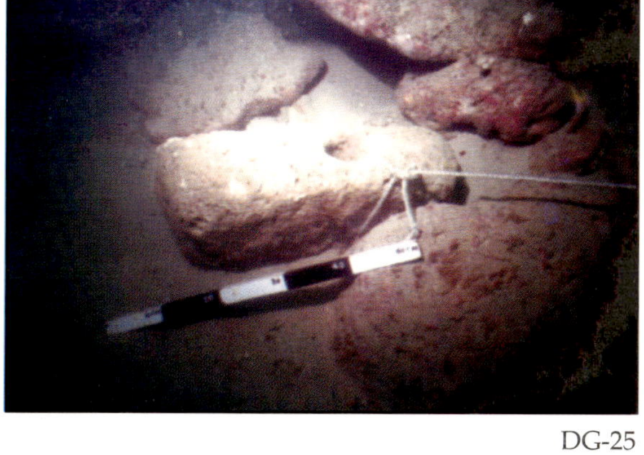

DG-25

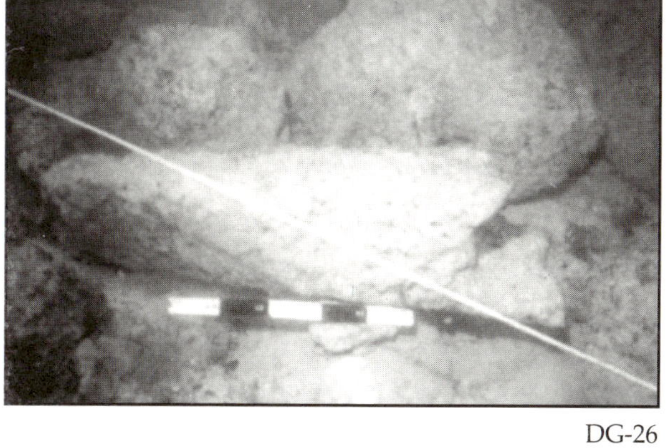

DG-26

Pl. 30. Indo-Arab type of stone anchors No. 19 to 26 from Dwarka waters.

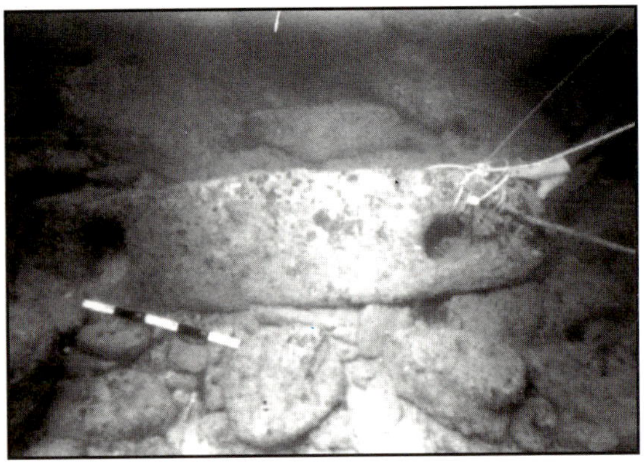

DG-27

DG-28

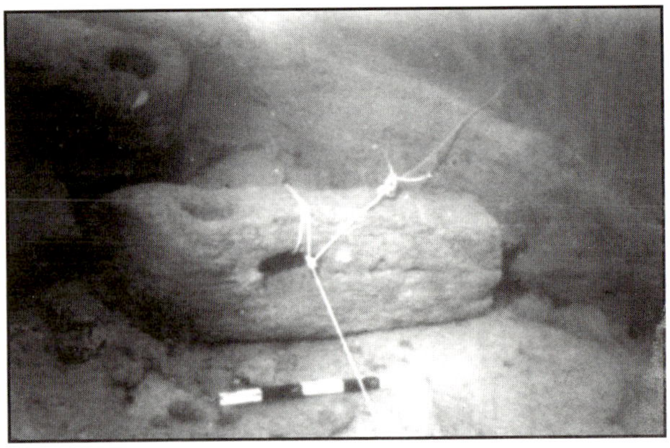

DG-29

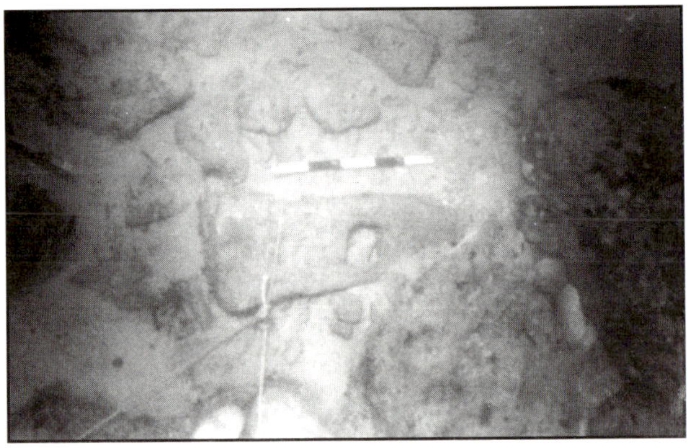

DG-30

DG-31

DG-32

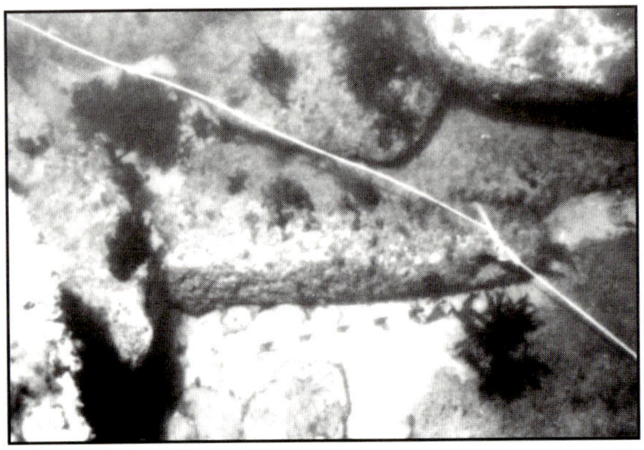

DG-33

DG-34

Pl. 31. Indo-Arab type of stone anchors No. 27 to 34 from Dwarka waters.

> the mighty struggle of thoughts, in the waters of the waves of lust and malice and dullness, not letting them divert it in one direction or another. This, O king, is the first quality of anchor he ought to have. And again, O king, as the anchor floats not, but sinks down, and even in waters a hundred cubits deep holds the ship fast, brings it to rest; just so, O king, should the strenuous Bhikshu, earnest in effort, when he receives support, and fame, and honour, and veneration and reverence, and offerings and praise, be not lifted up on the summit of the support of the fame, but keep his mind fixed on the idea of merely keeping his body alive. This, O king, is the second quality of the anchor he ought to have. For it was said, O king, by Sariputta, the Elder, the Commander of the Faith: As the anchor floats not, but sinks down beneath the waves, So be abased, not lifted up, by praise of gifts.

A 10th century Sanskrit text *Tilakamanjari* (Gopal, 1970) mentions about the stone anchor as "when a ship halted its heavy anchors made of rocks were lowered down." Further, the same text refers to ships being tied to strong wooden poles thrust deep into the earth. At the time of the departure of a ship the rocks which cut its speed were pulled up and the ropes that tied it to the pole were cut. The *Samraiccakaha* refers to the anchor being loosened to stop the drifting of the ship.

It will be interesting to take note of another 10th century AD story relating to stone anchors, mentioned by Hourani (1975) with reference to the *Kitab Ajaib al-Hind* (The Book of Wonders) written in the middle of the 10th century by Buzurg ibn-Shahriyar of Ramhurmuz, one of the greatest Arabic writers who collected many of his stories from captains at the Iranian port of Siraf. According to him, one of the captains on the China route commented as follows:

> Our ship was overloaded with heavy cargo from Siraf. When we reached the entrance of the China Sea, we saw Abharah, a man well-versed in the China Sea. After paying a huge price to Abharah, he came on our boat and ordered us to cut the larger mast, take up the anchors, and then to cut the cable of the larger anchor, which we did. Then he ordered us to cut the cable of other anchors and continued until we had thrown six anchors into the sea. We arrived in China and stayed there until we had sold and bought, repaired our ship, finally left for Siraf. While passing the same place where Abharah had boarded our ship he ordered us to drop anchor and send 15 men in the life-boat and bring our anchors from the reef. They brought a total number of three anchors.

Later Abharah explains that, "the total weight of the anchors on the boat was half of the entire cargo on board. Therefore to reduce the weight I had jettisoned the anchors. At that time it was high tide and now it was low tide so the anchors can be seen from the surface". From the above testimony it appears that the anchors were very heavy. These might be grapnel type anchors only. There are evidences that these anchors weighed more than 1 tonne. When the ships were overloaded, anchors were dropped at a known place and on the return voyage they were picked up again. In the process they lost some anchors at times.

The most interesting reference comes from *Periplus of Erythraean Sea* (Schoff, 1974) where the author specially mention the Gulf of Baraca (which is generally accepted as Dwarka and Bet Dwarka area). He says,

> the Gulf of Baraca contains 7 islands. Those who come to the entrance of this bay escape it by putting about a little and standing further out to sea; but those who are drawn inside into the Gulf of Baraca are lost; for the waves are high and very violent and the sea is tumultuous and foul, and has eddies and rushing whirlpools. The bottom is in some places abrupt, and in others rocky and sharp, so that the anchors lying there are parted, some being quickly cut off, and others chafing on the bottom.

4.2 STONE ANCHORS IN INDIAN ARCHAEOLOGY

Though it has been a well established fact that the oldest known civilization of the Indian subcontinent had overseas trade relations with the contemporary civilizations of the west, material remains in terms of shipwrecks, identification of trade routes and port structures are negligible. The excavations at Lothal in the head of the Gulf of Khambhat yielded a few large pierced stones which the excavator has identified as anchor stones (Rao, 1979). In contrast, a few scholars believe that these are counter-weight of water lifter from the tank which are still in use in many parts of Gujarat. The other instance of stone anchor from Bronze Age came from Kuntasi on the northern Saurashtra coast where a structure claimed to be a jetty was excavated (Dhavalikar *et al.*, 1996). It is very difficult to assess independently the uses of these stones in respect of navigation. However, underwater and coastal explorations of a few later period sites along the Indian coast yielded well-defined anchor stones. The prominent sites in Gujarat are Dwarka (Gaur *et al.*, 2001), Bet Dwarka (Gaur *et al.*, 2005), Somnath (Gaur *et al.*, 2002), Mithi Virdi (Gaur *et al.*, 2005), Miyani and Visawada (Gaur *et al.*, 2006), in Maharashtra, Vijaydurg (Sila *et al.*, 1998), Dabhol (Gaur *et al.*, 2007) and Sindhudurg (Sila and Gaur, 1997), in Goa waters (Sila *et al.*, 2003), Kerala coast (Sila *et al.*, 2005), and in Tamil Nadu coast (Athiyaman, 2001).

The underwater investigations in Dwarka water (in three seasons between 1998 and 2001) yielded more than 120 stone anchors found between the intertidal zone and 12 m water depth. There are varieties of stone anchors found from Dwarka and they have been broadly categorized into three three groups, namely, (i) composite type, (ii) Indo-Arab or grapnel type, and (iii) ring-stone type. They have been further categorized as per shape difference, the details which of are given below.

4.3 COMPOSITE STONE ANCHOR

Frost (1991) defines composite anchor as an anchor having more than one hole, wooden 'arms' struck into supplementary holes so as to protrude on either side and ensure that whichever way the stone falls one end of a stick will hook it into sand or mud. Triangular composite anchors are often carved from a flat, thin stone block, giving a triangular shape with a circular upper hole and two lower holes, which may be either rectangular or square. Sometimes, however, composite anchors may have one, two and four holes.

Out of the total 123 stone anchors observed, 36 are composite type anchors. Table 1 to 4 provide the details of their weight and dimensions along with other specific details of holes. On the basis of their shapes and perforation, the anchors have been broadly divided into three groups. Group 1 comprises the anchor with all circular holes and irregular shape. The anchors of Group 2 are perfectly shaped with a circular upper hole and the lower two holes are square or rectangular. Anchors of Group 3 are characterized by the introduction of an additional hole in the middle or upper side. The details of each group are given below.

4.3.1 Group 1

The anchor nos. 2, 8-11, 17, 18, 24, and 31 are included in this group. The main characteristic features of this group of anchors are (i) oval-shaped and some are irregular in shape, (ii) all holes are circular, (iii) number of holes varied from 1 to 3, and (iv) inconsistency in sizes of holes. Anchor no. 2 (laterite with rough surface) and 11 (limestone with smooth surface) are oval in shape and are well preserved. They have blunt edges, thin section and smooth surface and rope marks are also noticed. The sizes of holes are consistent. Often the upper hole is wider than the lower holes. Majority of the anchors are made of limestone, which is locally available. Other anchors in this group are irregular in shape having a rough and porous surface. Holes of these anchors are not properly shaped and two of them have only one hole (24 and 31). The weight of irregular-shaped anchors varied from 22 to 114 kg. A few anchors were observed cemented with the seabed in Dwarka waters such as anchor no. 17, indicating perhaps they were lying since long on the seabed.

4.3.2 Group 2

A large number of composite type anchors (nos. 1, 3-7, 12-16, 21-23, 25-26, 28-30, 32-36) fall in this group. The important features of this group are (i) perfectly shaped, upper width is shorter than the lower one, forming a trapezoid shape, (ii) upper hole is circular whereas the lower two holes are either square or rectangular, and (iii) made out of limestone. Anchor nos. 3, 12, 26, 32, 34, 35 and 36 have trapezoidal shape. They have sharp edges and a smooth surface while the anchors with semi-circular top have blunt edges and a smooth surface. Upper portion of a few anchors (nos. 4, 5, 14, 16 and 21) in this group are broken; thus, it is difficult to determine their shape on the upper side. The weight of this group of anchors varied from 50 kg (no. 13) to 496 kg (no. 32). There is no proportional consistency between upper circular hole and lower square/ rectangular holes. However, in many cases, the diameter of the upper circular hole is greater than the length and width of the lower holes. All anchors are made out of limestone.

4.3.3 Group 3

Anchors with more than three holes are included in this group (anchor nos. 19, 20, and 27). The anchor no. 19 with more than three holes was recorded first from Dwarka waters. Interestingly, three rectangular holes exist and the fourth hole was circular which has been broken and placed at the upper portion. Anchor no. 20 has two holes – one circular and the other square. The holes are randomly made without any proper orientation. The upper circular hole is at usual position while an additional circular hole is placed in the middle of one side. The lower square holes are not in proper line. Anchor no. 27 has all the four holes circular. Only the upper hole is broken and the diameter of holes is very less. The remaining three holes form a triangle in the middle. The anchor has even thickness and blunt edges.

4.3.4 Discussions

Majority of the 'composite' anchors are made out of locally available calcareous rock, limestone from Okhamandal area. The anchors generally have a smooth and even surface. They usually have an upper circular hole and the lower two holes are either circular or rectangular/square. The upper hole is referred to as rope hole and the lower ones as fluke hole. In case of the anchors in Dwarka, Bet Dwarka and Somnath, the upper hole generally does not have any rope marks except a few which suggest that the upper hole will have a wooden log with circular section and both sides of equal length, and the rope would be tied with this wooden log. This will have two advantages, namely, (i) whatever way the anchor falls on the seabed, the upper portion of the anchor would touch the seabed due to wooden log, which will increase the grip of the lower wooden flukes on the seabed; and (ii) there will be less chances of trapping of the anchor in the rock since one side is always above the seabed. Raban (2000) has suggested that the composite anchors from the Indian Ocean countries might be having metal flukes in the lower holes; however, in the present case, none of the anchor holes had any metallic fluke. Though wooden flukes were also not found, however, it is possible that wooden parts would have floated away after breaking or disintegrated with the passage of time. A 17th-century shipwreck in Goa waters does not have a single piece of wood (Sila *et al.*, 2001), while the stone anchors are much older. A few anchors from the Mediterranean Sea have the remains of a wooden fluke dating back to the 14th century AD (Grossmann and Kingsley, 1996). However, it should be remembered that the Indian waters, being in tropical conditions with high productivity, could cause deterioation of wood much faster than the Mediterranean Sea.

The anchors with all circular holes are smaller in size and might have been used in small boats only. However, they appear to be of primitive stage. Anchor no. 2 is comparatively heavier than the other anchors with all circular holes and also has rope marks which suggest that initially rope was tied directly to the upper hole and wooden flukes with both pointed end of circular section were used in the lower holes. Anchors with two square/rectangular holes perhaps represent the second stage of the development of anchors, which facilitated better seabed gripping. Composite anchors with more than three holes might have been the next stage of the development. In these anchors, the additional hole is circular and often in the middle. Composite anchors have relatively less weight on the upper side; so, when the anchor is trapped in the seabed, there are chances of loosening the lower part of the anchor. However, additional hole in the middle will probably have a thick wooden log (as this hole generally is wider than the upper one) and a supportive rope would be tied with it, which will lessen the chances of loosening the anchors. Therefore, it appears that after some experience in navigation, an additional hole in the middle was introduced for the safety of the anchors. Interestingly, only two anchors with additional holes have been recorded so far. The reason for such less findings of these anchors could be due to the fact that a few anchors were lost in the sea or they were used rarely. One of the heaviest anchors discovered from Dwarka weighs about 496 kilogram. For comparison, Frost (1991) has reported an anchor weighing more than 600 kg from Ras Shamra.

4.3.4.1 Origin of the Composite Anchors

The stone anchors have been recorded with the Harappan context at Lothal (Rao, 1979) and at Kuntasi (Dhavalikar *et al.*, 1996), but they are of a different type and do not match with the anchors from Dwarka and Somnath. Significantly, they indicate that the use of stone anchors with a proper shape and size was prevalent in India from a few centuries earlier than the Christian era. The composite anchors from the Mediterranean Sea have been found associated with the Early Bronze Age having a single upper circular hole (Frost, 1970) and from Sozopol with three circular holes dating back to the 8th century BC (Dimitrov, 1976 and 1977), and perhaps the use of composite anchors continued in the Mediterranean Sea till very recently (Frost, 1979; Segal and Carmi, 1996). It is interesting to note that several composite anchors have been recovered along with amphorae sherds and lead anchors off Bet Dwarka, which could be dated to the early centuries of the Christian era (Sundaresh *et al.*, 2002). Though one can always be inclined to think that the introduction of composite anchor was by the Roman mariners along the Indian coast, the fact remains that findings of such triangular anchors are primarily restricted to the northwest coast of India, whereas the major sites of the Roman settlements were on the southern peninsula from where composite anchors have not been reported so far. Perhaps composite anchors were introduced in India by the early Arab mariners as they were trading with the Mediterranean countries from the early centuries of the Christian era and soon they were adopted by the local mariners as plenty of sources of raw material (limestone) is available along the Gujarat coast. This is also supported by the presence of a large number of stone anchors in Oman waters (Vosmer). There is every possibility of the existence of many anchorages in the Persian Gulf, which can be traced with the help of underwater archaeology. Subsequently, composite anchors remained in massive use in Gujarat coast till very recently.

4.3.4.2 Dating of Composite Stone Anchors

Though literary evidences suggest some indication of the use of stone anchors that provide some indirect evidence of the period during which they were prevalent, their dating is not an easy task. At the first instance, the anchors under discussion have not been found associated with land excavations where these can be correlated with other archaeological remains. Secondly, the locations where these anchors have been found in India, i.e., Dwarka, Aramda, Bet Dwarka,

Sindhudurg and Vijaydurg, suggest that they cannot be earlier than the Early Historic Period as the antiquities of these sites are not older than the Early Historic Period except at Bet Dwarka. These ports were very active during the Historical and the Medieval periods. As discussed earlier, a few stone anchors of the Indus Civilization of the 3rd millennium BC are different from those reported from underwater sites at Dwarka and Somnath. Therefore, the anchors under discussion cannot be associated with the Indus Civilization. Initially, the dating of anchors from Dwarka was based on the comparative study between stone anchors from Syria and Cyprus and dated to the 14th century BC (Rao, 1987). Recently, a hoard of 26 similar type of anchors have been reported from Qalhat in Oman (Vosmer) along with the pottery of early Islamic (7th century AD) to the Medieval Period (14th century AD). Based on the above discussion, the date of the composite anchors can be bracketed between the Early Historical to the Medieval Period at Dwarka and Somnath.

4.3.4.3 Description of Composite Anchors (Figs. 10-12; Pls. 24-27)

1. Composite anchor with semicircular apex and made of limestone. It has a smooth surface, blunt edges and even thickness. The anchor has a circular upper hole and the lower two holes are square/rectangular. A thin layer of marine growth is also noticed.
2. The anchor is pear-shaped with three holes in the centre forming an equilateral triangle and may be described as primitive type as all the holes are round; it has a rough surface

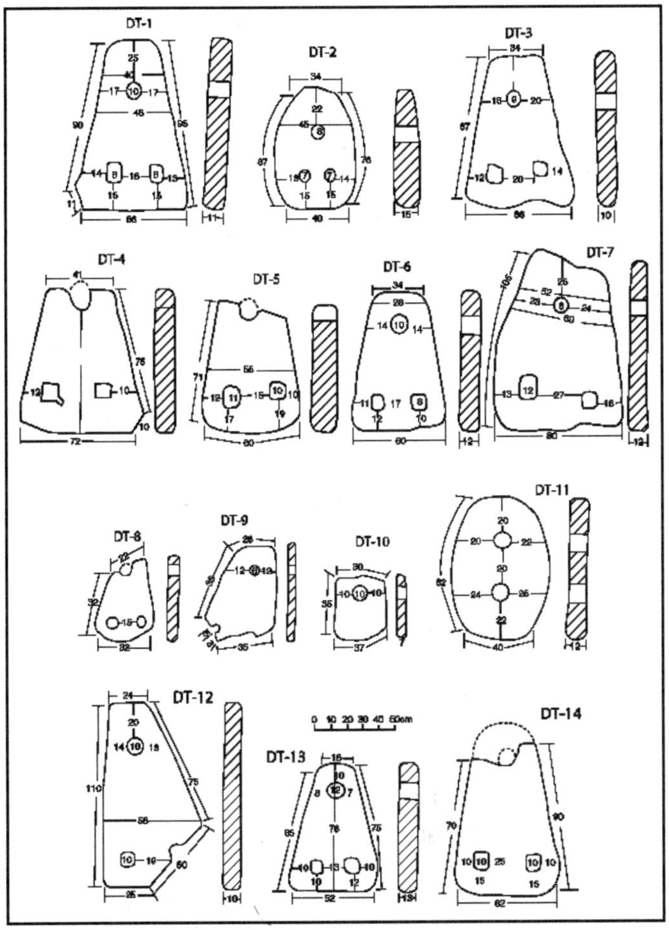

Fig. 10. Composite stone anchors No. 1 to 14 from Dwarka waters.

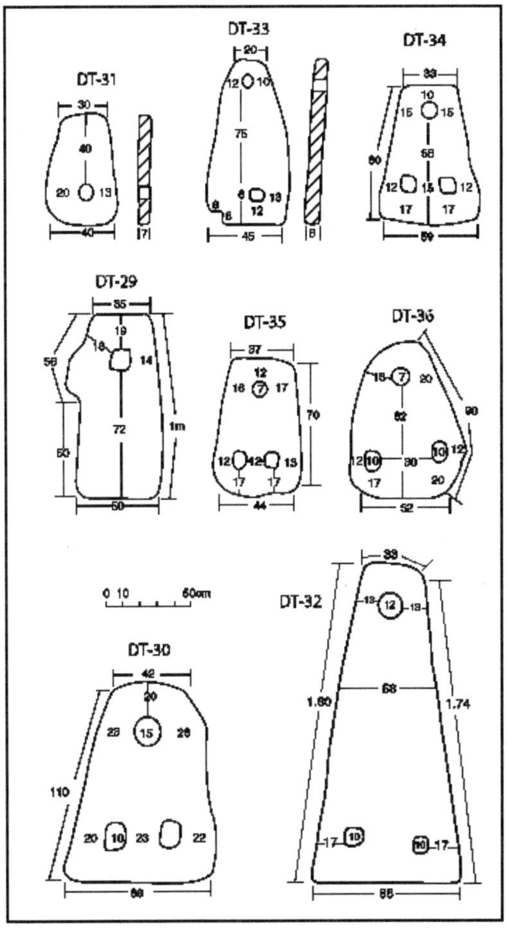

Fig. 12. Composite stone anchors No. 29 to 36 from Dwarka waters.

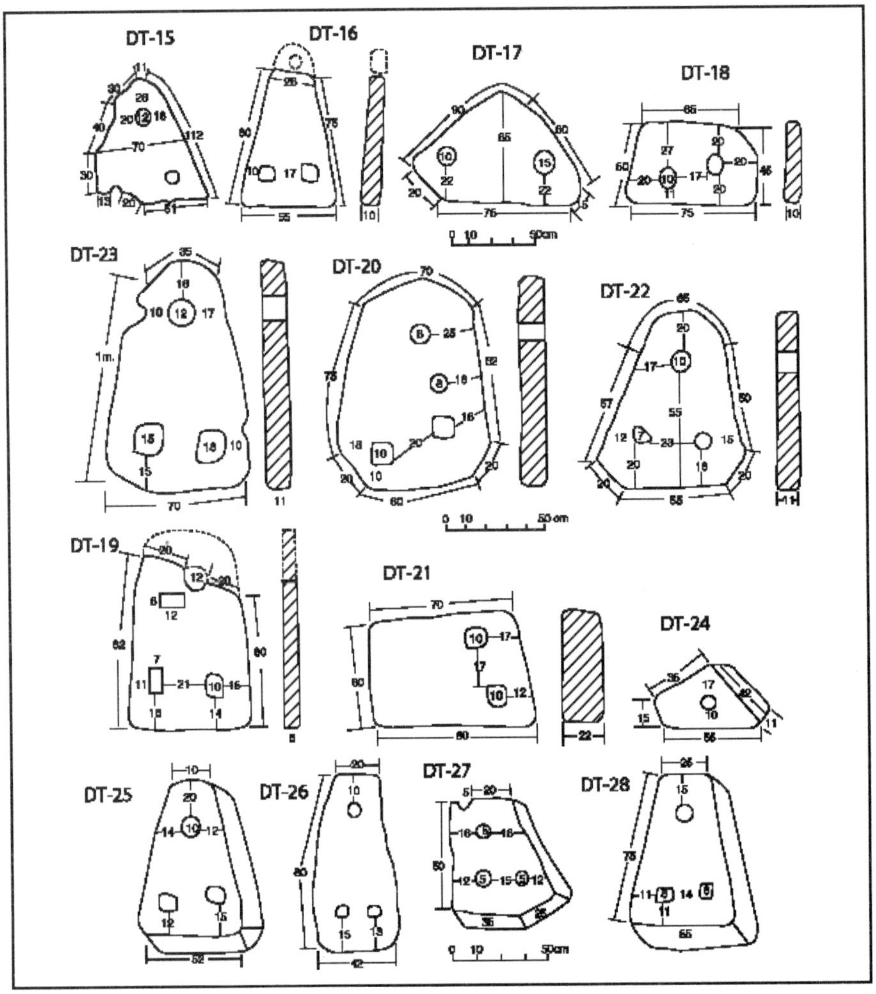

Fig. 11. Composite stone anchors
No. 15 to 28 from Dwarka waters.

and its edges are blunt. Rope marks were noted around the upper hole. It is only 20 cm thick but weighs 91 kg. The material of this anchor is laterite and that is not available locally.

3. Trapezoidal-shaped composite anchor has been chiselled perfectly. It has a smooth surface and sharp edges. The anchor has an upper circular hole and the two lower holes are square.
4. The broken composite anchor is chiselled out from limestone. The upper portion of the anchor along with a half upper hole has been broken and the lower two square holes are preserved. The anchor has a smooth surface and sharp edges with even thickness.
5. The broken composite anchor is made out of limestone. The upper portion with a half circular hole is broken. The thickness of the anchor increases gradually towards its lower side. It has a rough surface and blunt edges. It has an upper circular hole and the lower two holes are square.
6. Another stone anchor with a semicircular apex was retrieved earlier. The thickness of stone gradually increases on the upper side. It has a rough surface and blunt edges. The anchor is made of limestone. It has an upper circular hole and the lower ones are square.

7. This is one of the biggest triangular anchors, with an uneven thickness and rough surface. It has an upper circular hole and two square holes below. The material seems to be beach rock, which is locally available at Dwarka.
8. The small-sized anchor has three circular holes. The upper portion with a broken hole. It is made of limestone. It has a smooth surface and blunt edges.
9. The small-sized anchor has been cut from a thin limestone block. Only the upper hole is surviving while the lower ones are partially broken. The anchor has a pitted and rough surface with blunt edges.
10. Small-sized and partially buried anchor is survived with a wide circular upper hole. It has a thin section and is made of limestone. The anchor has a smooth surface and sharp edges.
11. Oval-shaped anchor has two circular holes. It is made of limestone. The anchor has a rough surface and blunt edges. It is the sole example from Dwarka waters.
12. Broken triangular anchor with an uneven and smooth surface. It is made of limestone. The anchor has an upper circular hole and the lower one is square. It has a rough surface and blunt edges.
13. Perfect triangular-shaped anchor is made of limestone. It has a porous surface. The upper hole has been widened due to extensive use. It has a rough surface and blunt edges. The lower holes are square.
14. Partially exposed stone anchor was noticed in the sandy seabed. The upper half portion including the broken circular hole; however, the lower two holes are intact. It is made out of limestone. The anchor has a smooth surface and blunt edges.
15. The broken triangular is made of limestone and it lay on a rocky beach. It has two square holes on the lower side and has a smooth surface, texture and even thickness.
16. The broken triangular anchor was found lying in a sandy channel at a water depth of 7 m. It is triangular in shape with two square holes and made of locally available limestone.
17. The broken triangular anchor is found at a depth of 5 m. The upper portion was broken but the two square holes at the base are intact. It was trapped between the rocks and is made of locally available limestone.
18. Two-holed triangular anchor was lying in a sandy channel at a depth of 7 m. Only the lower holes are preserved and both are circular in a shape. It has a very smooth and even surface and is made of sandstone.
19. The anchor was lying exposed on the seabed; it is made of yellow sandstone. The anchor has two square holes at lower and a rectangular upper hole. This is the first instance from Dwarka waters. It seems that one circular hole was broken following which they had made an additional rectangular hole. The anchor is lying at a water depth of 13 m.
20. The anchor was lying on the rocky seabed. There are unusually two circular upper holes and two square holes at lower side. It is difficult to presume the function of the two round holes together, may be precautionary measures have been taken. The anchor is made out of calcareous rock, which is rough and porous.
21. The anchor was observed about 10 m parallel to the cliff towards the shore from anchor 18 on a sandy bottom. The upper hole is broken while the lower two square holes are intact. This is made out of locally available calcareous rock and is lying in 14 m water depth.
22. The anchor was noticed close to anchor 20. It is a finely dressed anchor but the raw material is porous with a rough surface. It has an upper circular hole and two square holes on the lower side. The anchor is made of limestone.

23. This anchor was lying in-between two rocks near a grapnel anchor. It has a circular upper hole and two square holes below. This is made of calcareous rock. All the three holes have been eroded significantly. The anchor is lying in a water depth of 13 m.
24. Triangular anchor lying exposed on the rocky seabed. The lower portion of the anchor is broken. A small upper circular hole was noticed. It has a smooth surface and blunt edges. It is made out of limestone.
25. Triangular anchor was lying partially buried in sand. This is made out of limestone. The anchor has been eroded and the surface is porous and rough. It has blunt edges. An upper circular hole and two square holes on the lower side form a triangle. The anchor has an even thickness.
26. Triangular anchor was lying exposed on the sandy seabed; however it was surrounded with rock boulders. The anchor is made out of sandstone. It is one of the finest dressed anchors found in Dwarka waters. The anchor has a very smooth surface and blunt edges. It has a small upper hole and the two lower holes are square.
27. Triangular anchor made of limestone was lying exposed over sandy bottom and surrounded by small pebbles. It has an even thickness and the surface is smooth but the edges are blunt. The anchor has all three circular holes.
28. Triangular anchor made of limestone was lying exposed on the disturbed rocky seabed. The anchor has been shaped perfectly. It has blunt edges. Rope marks were noticed on the upper hole. A layer of pinkish marine and seaweed growth was also noticed.
29. The broken anchor was lying exposed on the sandy seabed. The anchor has only one lower square hole. It has a smooth surface and sharp edges. It is made out of limestone.
30. A triangular anchor is lying partially buried in sediment. It has an upper circular hole and the lower two holes are square. It has a rough surface and blunt edges.
31. Oval-shaped single holed stone anchor made out of limestone. It may be a natural one; however, it could have been used as an anchor for a small boat. It has only one upper circular hole and has a layer of pinkish marine growth. The anchor has a smooth surface and blunt edges.
32. Triangular anchor was lying partially buried under sediment. This is one of the biggest triangular anchors found so far from Dwarka waters. It has been cut perfectly. Though its surface is uneven, it is smooth and has sharp edges. The exposed portion is covered with a layer of pinkish marine growth.
33. Triangular anchor was lying partially buried in sediment. It is made out of limestone. As usual, it has an upper circular hole and the lower two holes are rectangular. It has a smooth surface and sharp edges. The upper portion of the anchor is perfectly semicircular. A layer of pinkish marine growth and seaweed growth was also noticed.
34. Triangular anchor made out of limestone was lying exposed over the rocky seabed. It has been cut perfectly. Its upper end is perfectly semicircular. It has an even surface and thickness. The anchor has a rough surface and blunt edges. As usual it has an upper circular hole and the lower two holes are square.
35. Triangular anchor cut out of limestone was lying exposed on the rocky seabed. One lower hole has been broken. The upper portion is semicircular. It has a smooth surface, blunt edges and even thickness. As usual it has an upper circular hole and the lower two are square/rectangular.
36. Triangular anchor was lying exposed over the sandy seabed in a channel. It has been perfectly chiselled. The anchor has a smooth surface and blunt edges. The anchor is located in the western most direction towards the lighthouse.

TABLE 1: COMPOSITE STONE ANCHORS DISCOVERED FROM DWARKA WATERS
(ALL MEASUREMENTS ARE IN CENTIMETRE)

Sl. no.	Location and Year	Raw material	Weight (esti-mated)	Length	Width Max.	Width Min.	Thick-ness	Dia. of top hole	Lower holes
1	Before 1997	Limestone	87*	95	66	32	11	10	8 x 9, 9 x 9
2	-do-	Laterite	91*	70	52	20	15	8	7,7
3	-do-	Limestone		87	66	34	10	9	10 x 10,10 x 10
4	-do-	Limestone	80	79	72	41	12	8	12 x 12, 11 x 9
5	-do-	Limestone	82*	71	60	39	18	10	11 x 11,10 x 10
6	-do-	Limestone	78*	76	60	34	12	10	8 x 8, 9 x 8
7	-do-	Limestone	147	105	80	56	12	8	12 x 13,13 x 13
8	-do-	Limestone	16*	40	36	22	6	4	4,4
9	-do-	Limestone	18*	59	40	27	5	6	5
10	-do-	Limestone		35	37	30	7	10	—
11	-do-	Limestone	73	80	58	25	12	9	9
12	-do-	Limestone	69	110	58	24	10	10	10 x 10
13	-do-	Limestone	45*	76	52	18	13	12	9 x 9, 9 x 9
14	-do-	Limestone	105*	90	62	40	12	11	10 x 10,10 x 10
15	1998	Calcareous	120	112	70		12	12	10 x 10
16	1998	Calcareous	54	80	55	28	10		10 x 10,10 x 10
17	1998	Calcareous	114	65	80	40	18		15, 12
18	1998	Calcareous	58	50	75	65	10		10, 10
19	1999 UW	Sandstone	63	82	65	50	8	12 x 6	10 x 7, 10 x 8
20	1999 UW	Calcareous	109	100	75	45	11	88	10 x 10,10 x 10
21	1999 UW	Calcareous	154	80	62	50	22		10 x 10,10 x 10
22	1999 UW	Calcareous	74	85	54	30	11	10	7 x 7, 7 x 7
23	1999 UW	Calcareous	89	100	70	35	11	12	15 x 18, 5 x 15
24	2000 UW	Limestone	24	55	42	15	11	5	
25	2000 UW	Limestone	45	70	52	18	10	10	10 x 10, 10 x 10
26	2000 UW	Limestone	73	80	42	20	14	5	5 x 7, 5 x 7
27	2000 UW	Limestone	25	50	35	25	10	5	5, 5
28	2000 UW	Limestone	62	78	55	25	12	8	8 x 8, 8 x 8
29	2001 UW	-do-	83	90	50	35	11	-	11 x 12
30	2001 UW	-do-	209	110	88	42	17	15	14 x 11,10 x 10
31	2001 UW	-do-	33	60	40	30	7	8	
32	2001 UW	Limestone	496	180	86	33	27	12	10 x 11,10 x 10
33	2001 UW	Limestone	47	91	45	20	8	6	6 x 8, 6 x 8

34	2001 UW	Limestone	93	80	59	33	18	9	8 x 8, 7 x 7
35	2001 UW	Limestone	37	70	44	37	8	7	6 x 6, 6 x 8
36	2001 UW	Limestone	69	95	62	25	10	7	8 x 10, 8 x 10

* Actually valued

4.4 GRAPNEL OR INDO-ARAB TYPE STONE ANCHORS

Before the arrival of the Portuguese in the Indian waters, Arabs were the dominating power in the Indian Ocean to undertake and control maritime trade since the 2nd century AD. Initially, they ventured into the Mediterranean Sea and were involved in several raids (Hourani, 1975). For a long time they were the supreme maritime power in the Red Sea and occupied the East African Coast as well as the Indian Coast. They reached Far East up to China and several stories of the Arabian Nights have been narrated on the voyage from Arab to China. At several places on the Indian Coast, pirates were very active during this period. It is said that one had to be very careful during their voyage to Kachchh and Kathiawar. Hourani (1975:84) has mentioned that Arab navigation in the Indian Ocean underwent a slow decline since the entry of Vasco-de-Gama in the Indian Ocean.

4.4.1 A Brief history of Arabs in the Indian coast

In the 7th century the Arabs got control of the shipping in the Persian Gulf. In the mid-7th century they fought a war against Broach and Thana, possibly because they wanted to control their trade. In the beginning of the 8th century AD, the Arab traders began to trade with India by sea in great number, principally as traders and they had no serious ideas to conquest. As Strabo observed, "The Arabs are not very warlike even on land but rather traders and merchant and not at all warlike on the sea" (Sridharan, 1982). By the 9th century AD, Arabs had become so powerful that they were able to control the maritime trade route extending from the Red Sea to the south of China. But they were always threatened by pirates around Dwarka, Chaul, Dhabhol and Malabar coast. Muhammad bin Qasim made a peace accord with the people of Saurashtra who were fighting with sea pirates in Bet Dwarka (Chandra, 1977). After the 12th century AD, the Arabs controlled the east coast of India, especially after the decline of Cholas on the Coromandal coast.

A large number of grapnel type of anchors have been recovered from onshore, intertidal and offshore zones of Dwarka. However, the largest number of anchors has been reported from Dwarka totalling 63. Entire set of grapnel anchors have been divided into three groups on the basis of material and shape. Group 1 comprises of various sizes and the primary raw material is basalt. The anchors of Group 2 are made of fine limestone and they are very similar to Group 1 in shape. The raw material used for group 3 is also limestone but often porous and the upper hole is absent. The upper section of the anchor is circular. The characteristic features of these anchors is as follows:

4.4.2 Group 1

Six anchors fall in this group (nos. 5, 19, 20, 30, 58 and 60) and are made of basalt rock. The anchors are of medium size and heavy in weight when compared to the other groups. The maximum length of an anchor is 142 cm, which weighs 175 kg while the smallest anchor is 90 cm in length (lying underwater) and has an estimated weight of about 125 kg. Majority of the anchors have been chiselled from an elongated prismatic shape of a basaltic block. The anchors of this category have smooth surface and blunt edges except No. 5, which has a rough surface and the upper portion is irregular. Often the anchors have a thick broad section lower side and tapers upper side. Anchor no. 60 is made out of a circular elongated basalt block therefore the section at both ends of the anchor is circular. All anchors have an upper circular hole while the lower two holes are square or rectangular. A few anchors have chisel marks in the hole as well as on the surface.

4.4.3 Group 2

A large number (total 46 anchors) of stone anchors belong to this group. The basic material for this group is locally available limestone and other sedimentary rocks (Fig. 5b). The shape of the anchors is same as of Group 1. However, the sizes vary in length from 88 cm to 237 cm. Total 9 anchors (12, 13, 23, 27, 37, 38, 42, 49 and 56) have a length of more than 175 cm. Most of the anchors have smooth surface and sharp edges. A large number of anchors have been eroded. Some of the anchors also indicate extensive use as their holes have been widened. Often the anchors have an upper circular hole and the lower two holes are square/rectangular. Most of the anchors are lying exposed on the seabed. However, a few of them are lying partially buried in sand and a few trapped in-between rock boulders. The exposed portion of the anchors is covered with marine growth. A few anchors were also noticed in shallow waters, in less than 5 m under depth.

4.4.4 Group 3

Anchor nos. 4, 7, 11, 22, 26, 30, 31, 34, 36, 44 and 54 are included in this group. The basic material for this group is porous limestone. Often the section of upper end is circular and lacks upper hole. However, the lower two holes are usually rectangular/square. They have a smooth surface and blunt edges. It is suggested that this kind of stone might have been used as bollard. Most of the anchors lie exposed on the seabed and marine growth on the exposed portion was also noticed. The maximum length of the anchors in this group is 155 cm (no. 31) and the smallest anchor has a length of 68 cm (No. 7).

4.4.5 Discussions

The introduction of larger boats for long voyages across the sea necessitated the invention of a new type of stone anchor (different from the traditional composite type), which could hold the boat firmly in the sea. Generally, grapnel anchors are long, heavy, and made of fine raw materials compared to composite anchors of the Mediterranean Sea. Grapnel anchors are confined to the trade routes and ports visited by Arab traders around the Indian Ocean. Therefore, the credit to introduce grapnel stone anchor goes to the courageous Arab traders. The anchors have been modelled in such a way that gripping in sandy and rocky seabed is excellent and while lifting, it will come out even from a small passage. Often grapnel anchors are very strong and will not easily break while lifting or throwing in the sea. Another quality of this anchor is that it occupies lesser space on board as compared to other types. On several occasions, grapnel anchors were reused as mooring stones. Perhaps all these qualities led to the extensive use of this kind of anchor along the Indian Ocean. This is evident from the discovery of a large number of grapnel anchors from Dwarka and Bet Dwarka waters, Vijaydurg and Sindhudurg area along the Indian Coast.

4.4.5.1 Mooring or anchor stone?

Honor Frost (1979) terms these anchors as 'Arabic' and 'grapnel type'. However, Kapitan (1994:3) has recently argued, "anchor shanks with rectangular cross section cannot belong to grapnels.

Shaft of four-armed grapnels is square or circular". His arguments may be valid when we compare with modern iron grapnel type. However, the idea of introduction of these anchors in place of composite type ones suggest that somewhere the thoughts on grapnel type was hidden and even the function of these anchors could be like grapnel type. Further, Kapitan suggested that "the use of these anchors as stone shanks and they were not so much applied as ships anchors to be cast from on board, but served as mooring anchors". However, recent discoveries from Dwarka and Bet Dwarka do not confirm his hypothesis as out of the total 80 anchors, only 2 anchors have a length of more than 2 m and a maximum estimated weight less than 600 kg while a majority of anchors are between 1 m to 1.5 m long and have an estimated weight between 100 to 250 kg which can be lifted easily by 3-4 crewmen. From Dwarka, only two anchors can serve as mooring stone, and the rest (individually) cannot hold big boats like Arab Dhows. However, a few anchors of Group 3 might have served as mooring stones on jetties as they lack an upper circular hole and their upper section is circular which indicates that boats would have been tied with them but certainly they cannot serve as moorings on the seabed due to their low weight. Kapitan's (1994) other argument that around the Indian Ocean there is no safe harbour and they need an anchoring point and these anchors served as mooring anchors, is also not on firm footing as a number of places where these anchors are reported have safe harbour like Bet Dwarka and Vijaydurg on the west coast of India. There are also instances where these anchors have been reused as mooring stones on jetty and dockyard at Vijaydurg and Sidhudurg. One may be surprised to note that these anchors were also used as lintel stone in Vijaydurg Fort though this was not the principle function of these anchors.

4.4.5.2 Origin

Grapnel anchors have been found associated with Arab trade in the East African coast and in the Gulf countries. The various literature related to Arab navigation (Hourani, 1975) indicate that these anchors fundamentally modelled by Arab traders, later became popular along the Indian Coast. The description also suggests that a single boat used to have several anchors onboard. Each boat had a main anchor, which will be the heaviest among all (Hourani, 1975:115-16). This evidence has been recorded from Dwarka also, as at least nine grapnel anchors of more than 1.75 m in length were recorded (anchor nos. 12, 13, 23, 27, 37, 38, 42, 49 and 56) at different locations. At one place it is mentioned that the captain asked the crew to throw six anchors in the waters (Hourani, 1975:116). Moreland (1939:65) mentioned that "the anchors were, according to contemporary accounts, merely lumps of rock" and no mention of holes and sizes has been suggested. The lumps could be of two types of anchor: one may be ring stone and the other could be a grapnel anchor, which are under discussion in this paper.

4.4.5.3 Comparison with other sites in the Indian Ocean countries

A large number of Indo-Arab type of anchors have been reported from India, Sri Lanka, Maldives and the East African countries. Recent explorations at Malwan (Sila & Gaur, 1997) and Vijaydurg in the Sindhudurg district of Maharashtra yielded a number of stone anchors. In Vijaydurg, eight grapnel-type anchors used as lintels in the parapet wall of the fort wall during the Maratha Period have been identified (Sila *et al.*, 1998). This proves that these anchors were in use prior to the Maratha Period. Further, mooring stones have been identified on the unused dockyard located on the Vaghotan riverbank. Archaeologically, these ports are not dated earlier than the beginning of the Christian era.

The excavation of the 12th-14th century AD port at Periyapattinam in the Ramanathpuram district of Tamil Nadu yielded an identical grapnel three-holed mooring stone which was installed on the bank of the river Kappalaru. Further, oral tradition in the village clearly stated that the

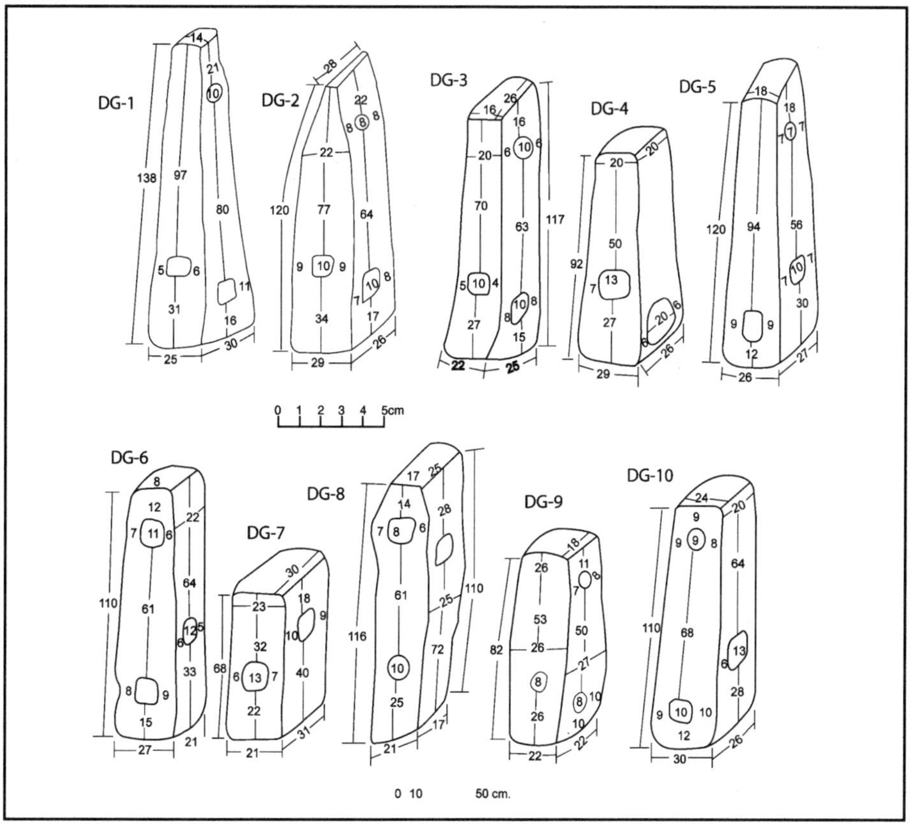

Fig. 13. Indo-Arab type stone anchors
No. 1 to 10 from Dwarka waters.

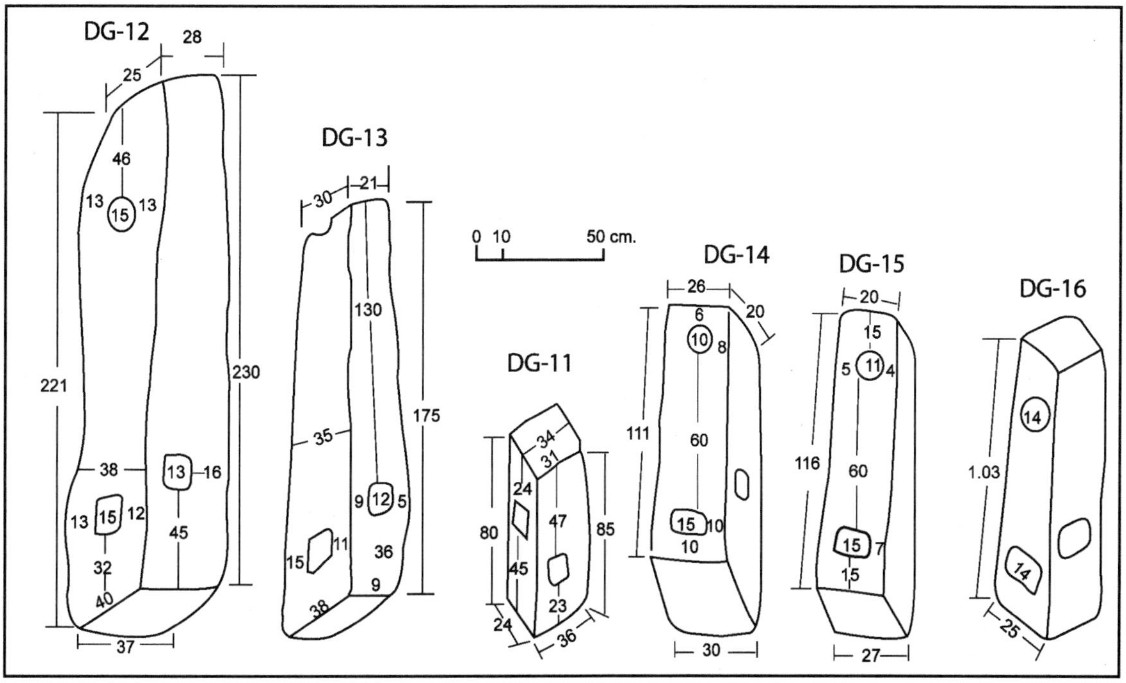

Fig. 14. Indo-Arab type stone anchors
No. 11 to 16 from Dwarka waters.

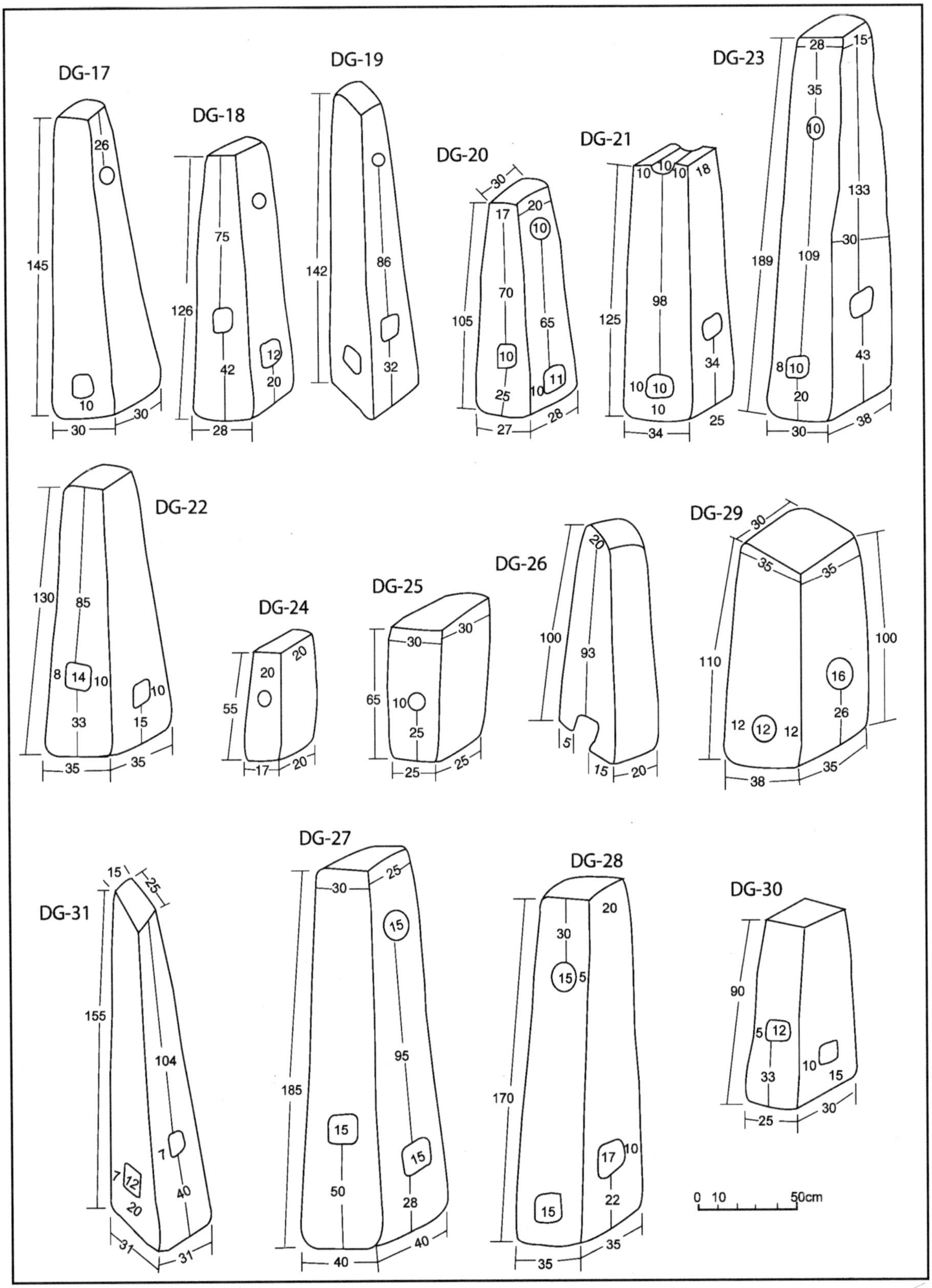

Fig. 15. Indo-Arab type stone anchors
No. 17 to 31 from Dwarka waters.

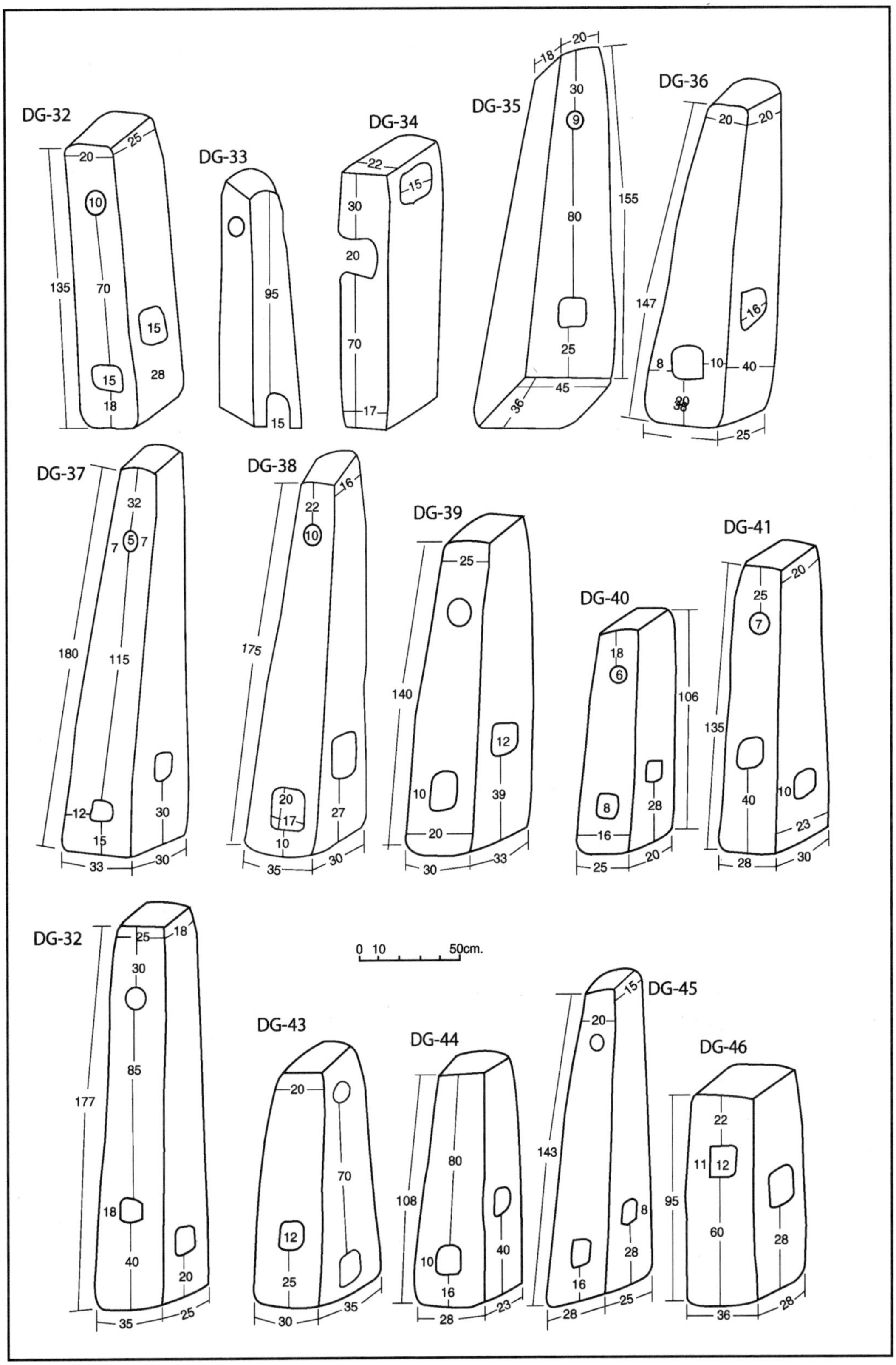

Fig. 16. Indo-Arab type stone anchors
No. 32 to 46 from Dwarka waters.

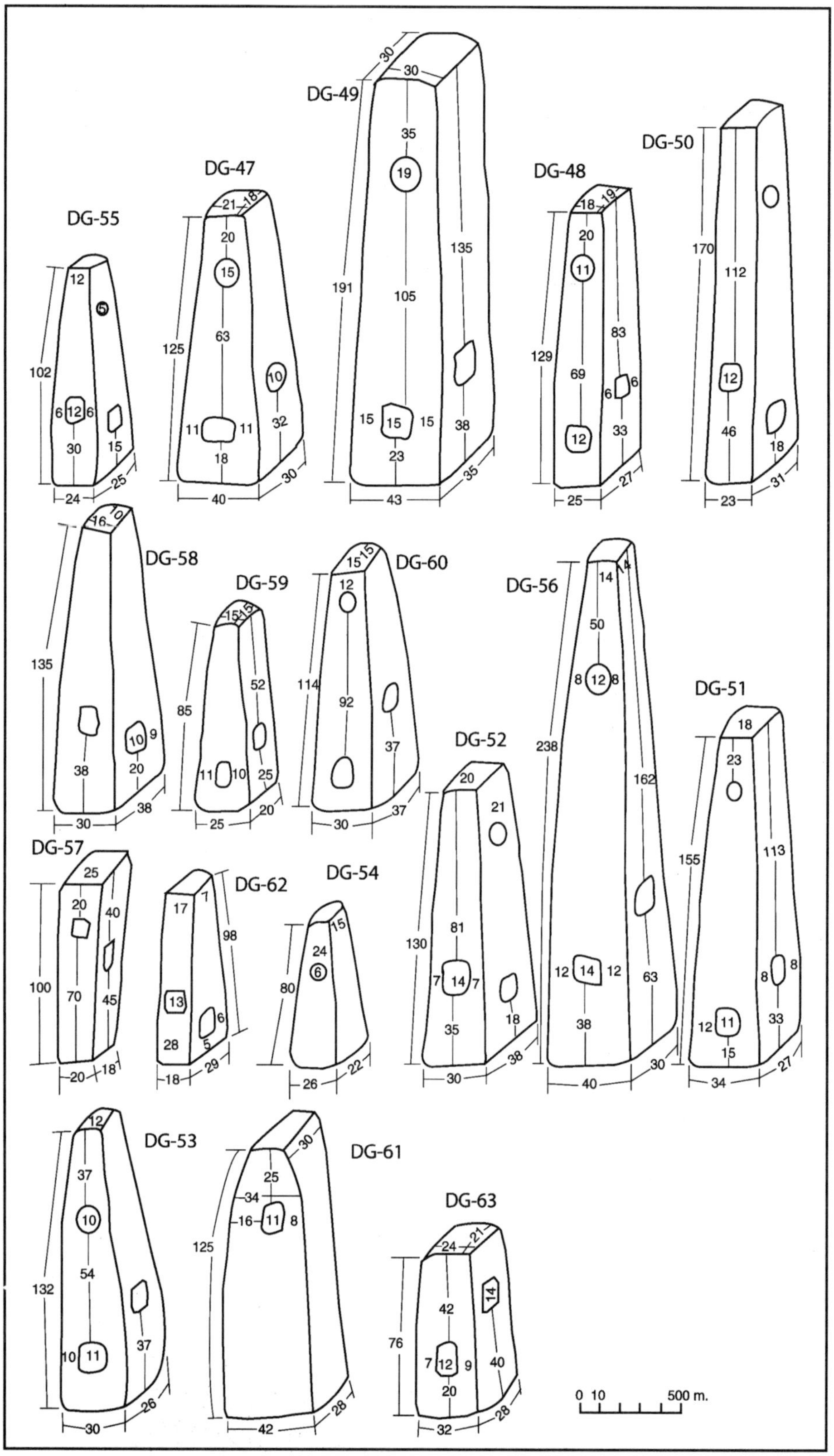

Fig. 17. Indo-Arab type stone anchors
No. 47 to 63 from Dwarka waters.

DG-35

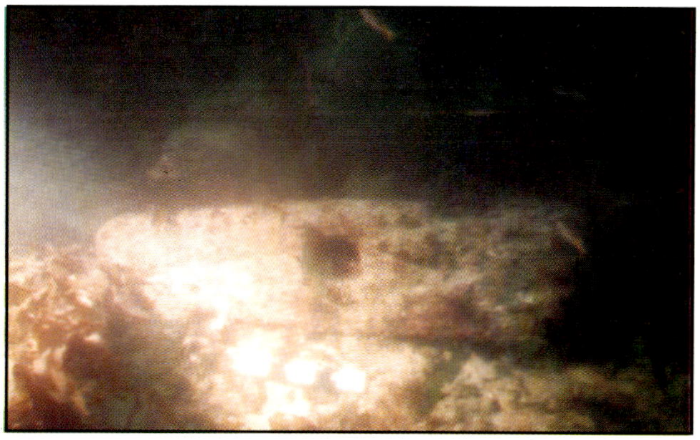

DG-36

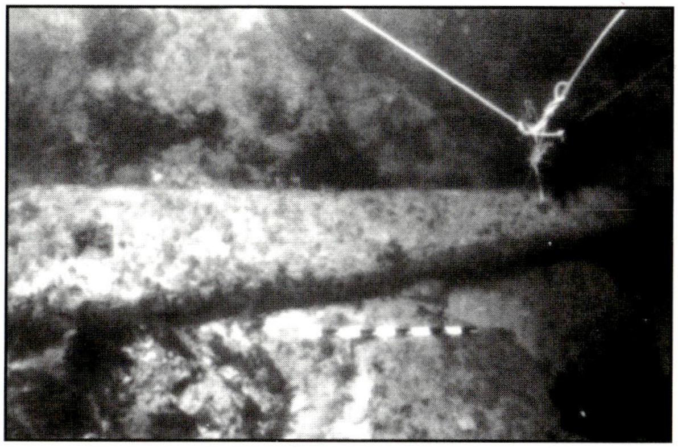

DG-37

DG-38

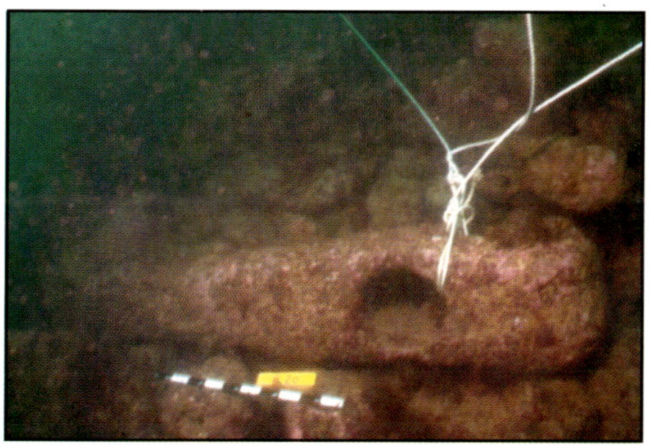

DG-39

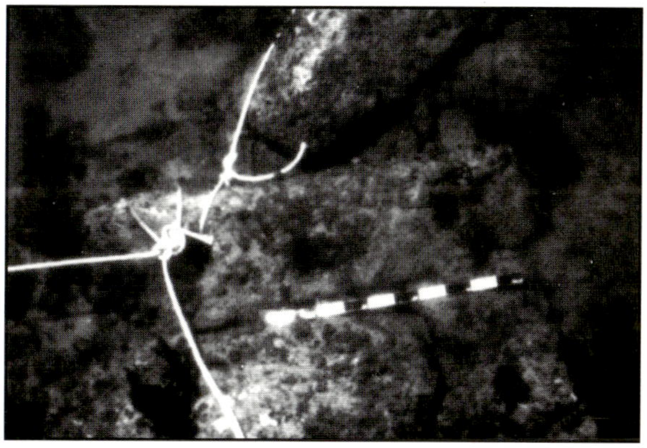

DG-40

DG-41

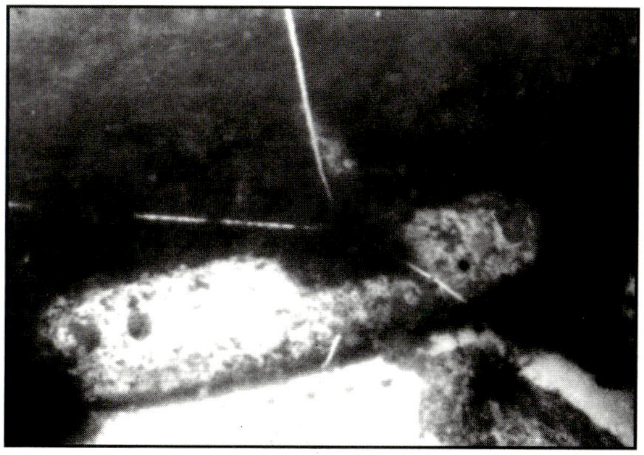

DG-42

Pl. 32. Indo-Arab type of stone anchors No. 35 to 42 from Dwarka waters.

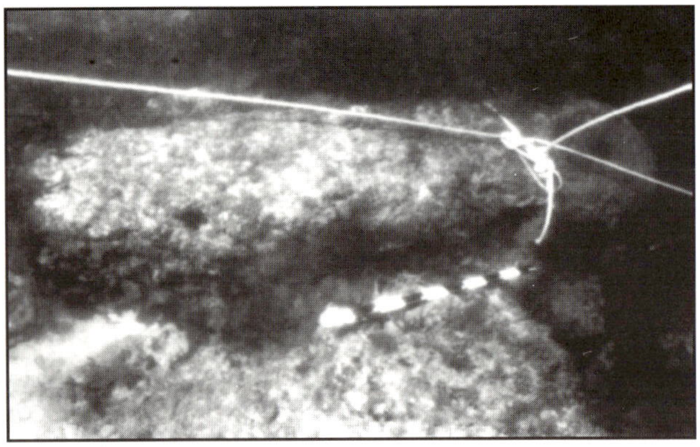

DG-43

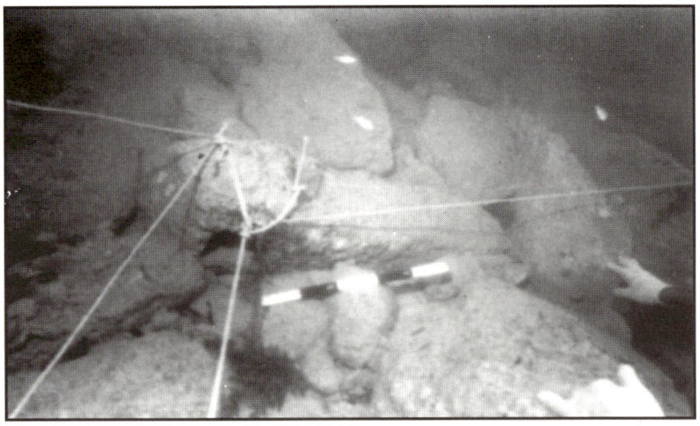

DG-44

DG-45

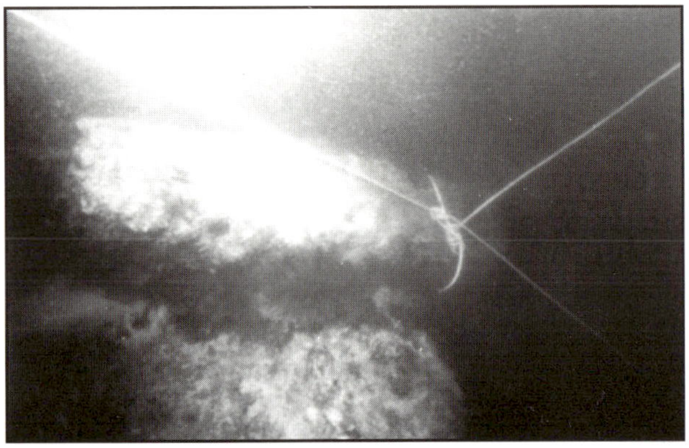

DG-46

DG-47

DG-48

DG-49

DG-50

Pl. 33. Indo-Arab type of stone anchors No. 43 to 50 from Dwarka waters.

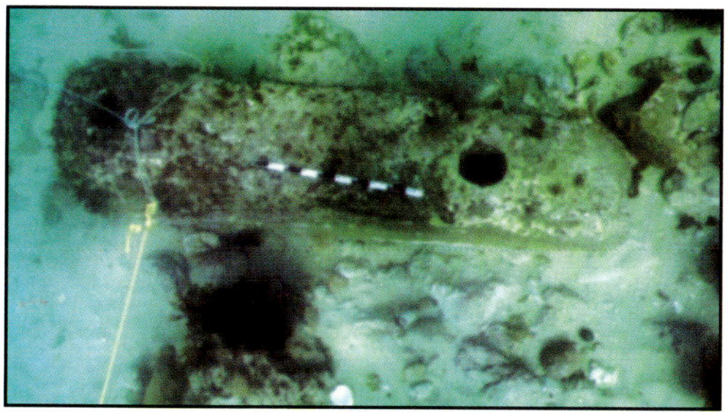

DG-51

DG-52

DG-53

DG-54

DG-55

DG-58

DG-56

DG-57

Pl. 34. Indo-Arab type of stone anchors No. 51 to 58 from Dwarka waters.

DG-59

DG-60

DG-61

DG-62

Pl. 35. Indo-Arab type of stone anchors No. 59 to 62 from Dwarka waters.

DR-1

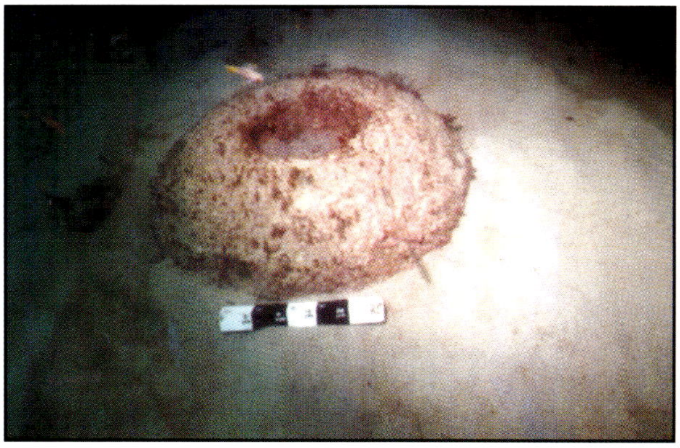

DR-2

DR-3

DR-4-6

DR-5

DR-7

DR-8

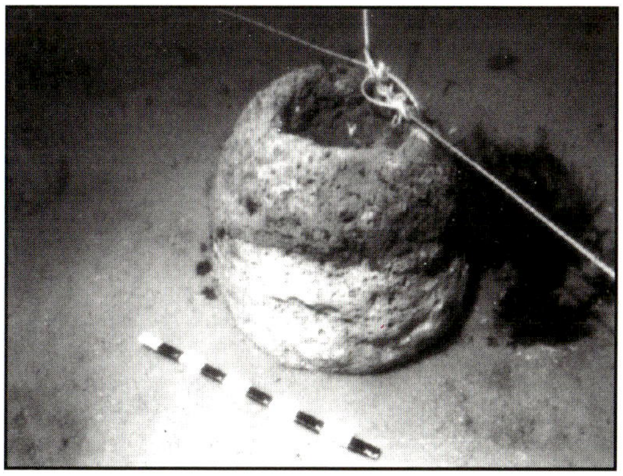

DR-10

Pl. 36. Ring stone anchors No. 1 to 10 from Dwarka waters.

DR-11

DR-12

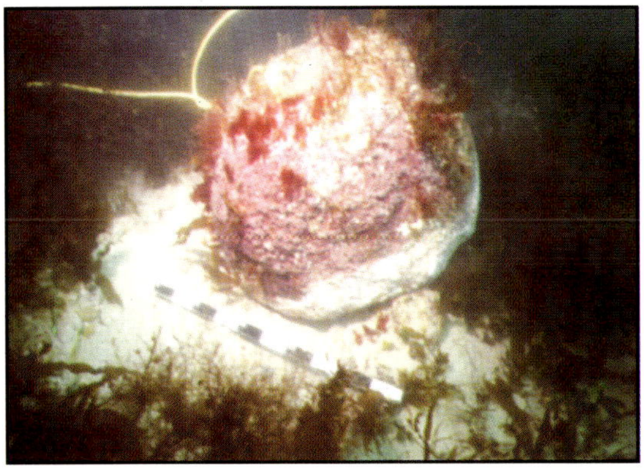

DR-13

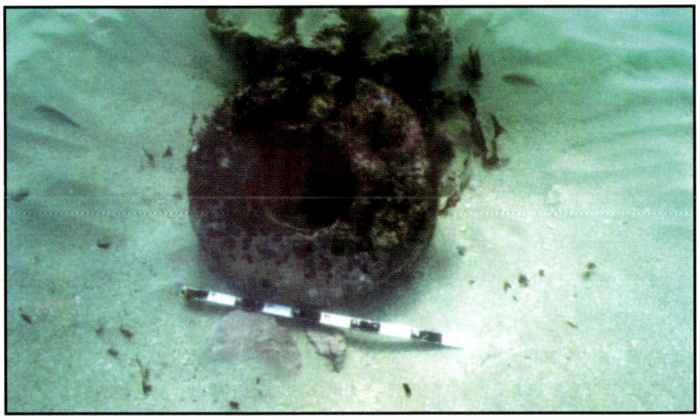

DR-14

DR-15

DR-16

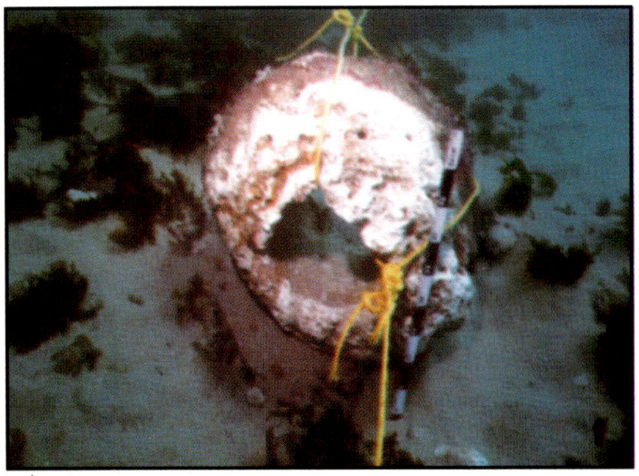

DR-17

DR-18

Pl. 37. Ring stone anchors No. 11 to 18 from Dwarka waters.

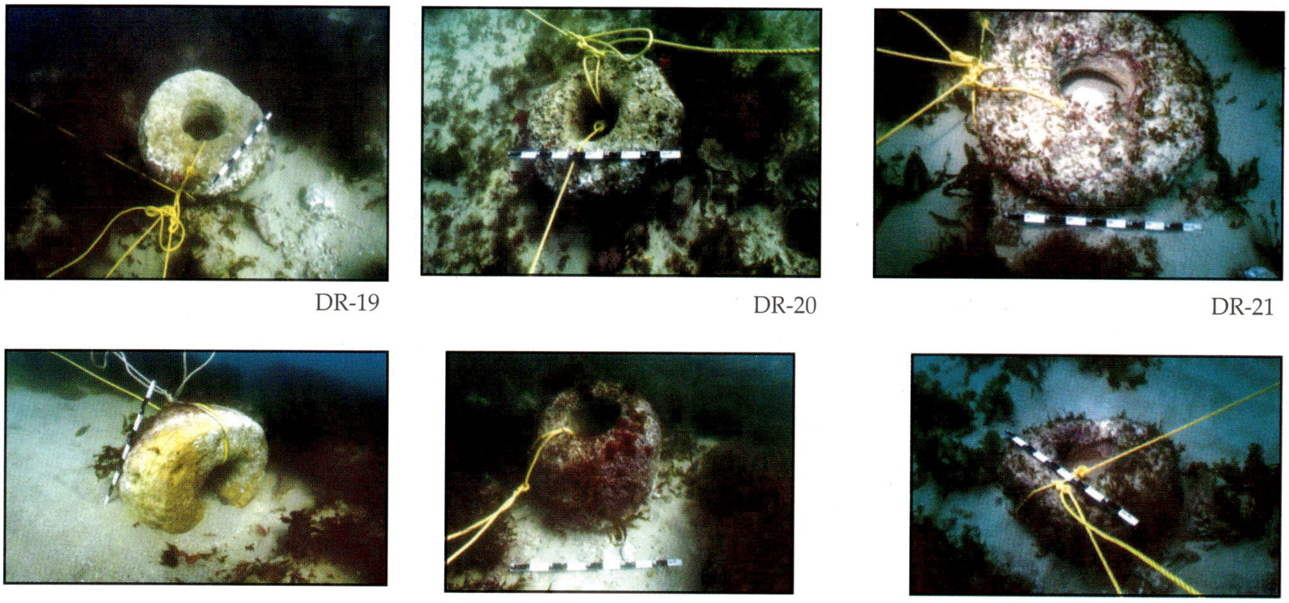

DR-19 DR-20 DR-21

DR-22 DR-23 DR-24

Pl. 38. Ring stone anchors No. 19 to 24 from Dwarka waters.

Boat motif no. 1 & 2.

Boat motif no. 3.

Pls. 39. & 40. Boat motifs on the hero stones from Aramda.

AT-1

Boat motif no. 4.

Pls. 41. & 42. Composite stone anchor from Aramda and boat motif on a hero stone near Positra.

stone was used for tying boats (Rajan, pers. comm.). Two more grapnels anchors from south of Tamil Nadu coast have been reported which are similar to those of Dwarka anchors (Athiyaman, 2001).

Raban (1990:303) believes that if these anchors were fundamentally made for safer anchorage in a coral bottom, typical of the Indian Ocean, then they were not necessarily confined to Arab boats or modelled by Arab influence. Parallels to this type of anchor have been found from the Red Sea on the Lone Mushroom wreck (Raban, 1990).

A similar type of grapnel anchor is reported from Kilwa Kisiwani and Mogadishu on the East African coast (Chittick, 1980:73-75). David Whitehouse's excavations at Siraf on the Persian Gulf have yielded two fragmentary stone shanks (anchors) of a similar nature, which he compares with a complete specimen found by the villagers at the same place (Whitehouse, 1970:141). According to his dating, these stratified fragments are not later than the 8th century.

4.4.5.4 Dating

It is unfortunate that not a single anchor has been found associated with stratigraphy. This makes dating of these anchors complicated. The anchors under discussion are entirely different from the Harappan anchors reported at Lothal (Rao, 1979) and Kuntasi (Dhavalikar *et al.*, 1996) and are therefore not comparable. As suggested earlier, these anchors are associated with the Arab traders who were active in the Indian Ocean countries from the 2nd century AD to the 15th century AD; therefore, a possible date of these anchors should fall between the 2nd century to the 15th century AD. However, along the Indian Coast, Arab traders were active after the 7th-8th century AD and grapnel anchors have been reported from the Medieval Period sites at Vijaydurg, Sindhudurg, and Tamil Nadu coast. The date of stone anchors from Tamil Nadu coast has been assigned to the 13-14th century AD (Athiyaman, 2001:1298). Onshore excavations in Dwarka and Bet Dwarka have suggested a huge habitational deposit belonging to the Medieval Period. The date of the grapnel anchors from Dwarka, Bet Dwarka and Somnath may fall between the 8th century AD and the 15th century AD. After the arrival of the western powers in the Indian Ocean, stone anchors were replaced with iron anchors. However, the other type of anchors (composite) were still popular among the fishermen on the Gujarat coast till recent times.

4.4.5.5 Description of Grapnel anchors (Figs. 13-17; Pls. 28-35)

1. Grapnel anchor made out of yellowish sandstone. Chisel marks surviving round the upper hole indicate less use of this anchor. The anchor has a rough surface and blunt edges.
2. The anchor is made out of limestone which is pinkish in colour. It has a circular upper section and lacks an upper hole; however, the lower square holes are surviving. The lowest hole has lost its original shape due to extensive use. The anchor has a rough surface and blunt edges.
3. The anchor is made out of calcareous rock. It has an upper circular hole and two rectangular holes at the lower side. The anchor has smooth surface and blunt edges and has even thickness.
4. Small anchor made out of limestone. It has a round upper portion without an upper hole, but has two rectangular holes below. It has a smooth surface and blunt edges. It is possible that this anchor was used as a bollard.
5. This is the most important grapnel anchor made out of basalt rock. It was found in Dwarka waters. As this material is not available locally, its origin should be sought

either in the lower Saurashtra region or out of the Gujarat coast. It has the usual an upper circular hole and two square holes at lower side. It tapers towards the top, has a smooth surface and blunt edges. Anchor no. 16 made of very fragile calcareous rock was found. The lower two holes have lost their original shape but can be seen to be rectangular. Its upper portion has become smooth and round.

6. Anchor made out of yellowish sandstone. As usual the anchor has a circular hole in the upper portion and two square holes in the lower side. Chisel marks surviving round the upper hole indicate less use of this anchor. The anchor has a rough surface and blunt edges.
7. Another stone anchor has two square holes, which indicate that it is a broken grapnel. It is made of calcareous rock. It has an uneven surface but sharp edges.
8. The anchor has been made from a limestone rectangular block. The anchor has an upper circular hole and the lower two holes are rectangular. It has a smooth surface and sharp edges. One lower hole has lost its original shape.
9. The anchor has been shaped from a limestone rectangular block. This is one of the finest finish stone anchor reported from Dwarka waters. All three holes are circular and this is the lone example from Dwarka.
10. This is made out of limestone and is surviving in good condition. It has a smooth surface and sharp edges. The anchor has an upper circular hole and the lower two holes are rectangular.
11. The broken grapnel anchor was located during low tide in intertidal zone on a rocky beach. It is made of limestone, has a very smooth surface texture and two square holes in the lower part.
12. Grapnel anchor with an uneven lower part. It has an upper circular hole and two square holes below. This is the largest and may be one of the heaviest anchors found in Dwarka. It has a rough surface and is made out of yellow sandstone.
13. Broken grapnel but heavy and of the same type as no. 12. It has a smooth surface and appears to be of limestone. Growth vegetation is also noticed.
14. Grapnel anchor lying exposed on a rocky bed and has a circular upper hole and two square holes below. It has been cut from a limestone block and has a very rough and uneven surface.
15. This anchor is found partially buried in a channel at a depth of 6 m. It also has a circular upper hole and two square holes below. It is made of yellow sandstone and the lower holes have eroded significantly, suggesting prolonged use.
16. Grapnel anchor recorded at a depth of 7 m. The anchor is trapped between rocks. Holes have been widened due to extensive use. It is made out of locally available limestone. The upper hole is circular but the lowest hole is eroded.
17. Grapnel anchor was found between rocks and it has practically integrated with bedrock. It has a circular upper hole and one square hole below. It is made of locally available limestone, having a rough surface and blunt edges.
18. The grapnel anchor was partially buried in sediment and was exposed by removing sand. It has a circular upper round hole and two square holes below. The one face of the anchor has several grooves. This anchor is made of limestone and lying in 6 m water depth.
19. This grapnel anchor was found while locating the previously known submerged cliff. It was found lying just below the cliff area. It has sharp edges and seems to have been used for a short period. It has a very small circular upper hole and two square holes at the bottom on either side of the anchor. This is made of black hard rock, probably basalt.

20. This anchor was found about 100 m westward of anchor no. 19. The topography of this area is represented by high rocky cliff and a submerged channel merging at this place. The upper hole is very small having 5 cm diameter and there are two square holes below. The anchor has very sharp edges. It is 1.2 m in length and is made out of basalt.
21. The anchor has half broken upper hole and the lower two holes are rectangular on either side. This is 125 cm in length and made out of locally available limestone. It was found lying on the rocky slope of the cliff area at a depth of 13 m.
22. The anchor was lying in-between the rocks. It has a very rough surface and chisel marks are visible. It lacks upper hole while two square holes were noticed on either side of the anchor at the lower side.
23. This grapnel anchor was lying near anchor no. 6. This is the longest anchor found so far in this area which is 188 cm long. It has a circular upper hole and has the remains of chisel marks. The lower two holes are rectangular and appear to be made of yellow sandstone.
24. The broken grapnel anchor was found about 3 m towards the cliff from anchor no. 22. It has 55 cm length and the lower portion is broken. A circular upper hole was noticed. It is made of limestone and found in 13 m water depth.
25. The broken grapnel anchor was noticed about 5 m away from the cliff-side. It is 65 cm long and only a circular upper hole is surviving. The lower portion of the anchor is broken. This is made of limestone and is lying in a water depth of 13 m.
26. The anchor was found in-between several fallen rocks. The lower portion of the anchor is broken and only one broken lower hole was noticed while it lacks an upper hole. The length of the anchor is 100 cm and is made of limestone. This is lying in 12 m water depth.
27. The anchor was lying on the other side of the submerged cliff. It is trapped in-between the rocks. It is very big in size and is also very heavy. It has a length of 185 cm. This anchor has a circular upper hole and there are two square holes below. It has a very rough surface and is made of limestone. This is one of the biggest anchors found in this area, lying in a water depth of 13 m.
28. The anchor was noticed on the northern bank of the submerged channel near a rock and about 15 m away from anchor no. 8 parallel to the cliff. It is also a very heavy anchor. It has a circular upper hole of 5 cm diameter. One each square and rectangular hole below on either side of the anchor was noticed. It has a very rough surface and is made of limestone. It is lying in a water depth of 14 m.
29. The anchor was lying exposed on the seabed. The upper portion of the anchor is broken and two square holes were noticed below. It has a very rough surface and is made out of limestone. It has a length of 110 cm and lying in 14 m water depth.
30. The grapnel anchor was found partially buried below the rock. The upper portion of the anchor is hidden under the rock. Lower holes are square and a length of 90 cm was measured of the visible portion. It is made of basalt rock. The anchor has very sharp edges and is lying in 12 m water depth.
31. The anchor was lying in shallow water and buried in sediment. It is a heavy anchor and has no upper hole. Two square holes were noticed at lower side. It is made of limestone. Vegetation growth was also noticed. The anchor is lying in a water depth of 6 m.
32. Grapnel anchor lying exposed on rocky seabed. It is made out of limestone. It has an upper circular hole and the lower two holes are rectangular. The anchor has a smooth surface and sharp edges.

33. Grapnel anchor was lying exposed on the rocky seabed. The lower portion of the anchor is broken and the upper portion has a circular hole and the section of the anchor have become round. It has a porous surface with blunt edges. It is made of limestone.
34. Grapnel anchor was lying exposed on the seabed. It is broken on the lower side and only one broken lower hole is surviving. It has no upper hole and the upper section of the anchor is circular. It has a rough surface and blunt edges. It is made out of limestone.
35. Grapnel anchor was lying exposed on the rocky seabed close to the above anchor. This is one of the heaviest anchors. It has an upper circular hole and the lower two holes are rectangular. It is made out of limestone. It has a rough surface and blunt edges.
36. Heavy grapnel anchor was lying exposed on the rocky seabed. It has an upper circular hole and the lower two holes are rectangular. The upper section of the anchor is round. It has a rough surface and blunt edges. The anchor is made out of limestone.
37. Grapnel type anchor was lying exposed on rocky seabed. This is a very heavy anchor. This is one of the finest finished anchor recorded from Dwarka waters. It has a smooth surface and sharp edges. The anchor has an upper circular hole and one each rectangular and square hole on the lower side. The anchor was trapped in-between the rocks. It is made out of sandstone.
38. Grapnel anchor was lying exposed in-between two rocks. This is a very heavy anchor and made out of limestone. It has an upper circular hole and the lower two holes are rectangular. The holes have been eroded significantly and widened. The surface of the anchor is rough and has blunt edges.
39. Grapnel type anchor was lying exposed on the rocky seabed. It is made out of locally available limestone. Due to erosion the surface of the anchor has become rough and has blunt edges. It has an upper circular hole and one each lower one is square and rectangular. Perhaps the lowest rectangular one has been widened and become rectangular while the original shape must have been square.
40. Grapnel type anchor was lying exposed on the seabed. It has a very smooth surface and blunt edges. It is made out of sandstone. The anchor has an upper circular hole and the lower two holes are square.
41. Grapnel type anchor was lying exposed on the rocky seabed. It is made out of sandstone. The anchor has a smooth surface and blunt edges. It has an upper circular hole and the lower two holes are square in shape.
42. Grapnel anchor was lying partially buried in sediment. It is made out of limestone. This is one of the heavy anchors. The surface is rough and has sharp edges. It has an upper circular hole and one each lower hole is square and rectangular.
43. Grapnel anchor was lying partially buried under the rock. It has a very rough surface and blunt edges. As usual the anchor has an upper circular hole and the lower two holes are rectangular. The anchor has been eroded. It is made of limestone.
44. Grapnel type anchor was lying exposed on the rocky seabed. It is made out of limestone. The upper circular hole is not present and the lower two rectangular holes are surviving. It has a rough surface and blunt edges.
45. Grapnel type anchor was lying on rocky seabed in shallow water. It is made out of limestone. The anchor has a rough surface and irregular shape. It has an upper circular hole and each lower hole is rectangular and square. It has blunt edges. Growth of seaweed is also noticed over the exposed portion.
46. Grapnel anchor was lying on rocky seabed in shallow waters. The upper portion of the anchor is broken and lower two rectangular holes are intact. It has a rough surface and irregular shape. It is made out of limestone. Seaweed growth is also observed.

47. Grapnel anchor made of limestone was lying exposed on disturbed rocky sea bed. It has a rough surface and sharp edges. The section of the upper and lower ends is square and rectangular respectively. A layer of pinkish marine growth and seaweed is noticed. The anchor has an upper circular hole and the lower two holes are square.
48. Grapnel type anchor was lying partially buried in sediment. It is made out of limestone. The anchor has a smooth surface and blunt edges. The upper side of the anchor has turned roundish. The exposed portion witnessed the growth of a thin layer of pinkish marine growth and seaweed growth is also noticed. As usual, it has an upper circular hole and the lower holes are square/rectangular.
49. Grapnel anchor made out of yellow sandstone. This is one of the biggest grapnel anchors found in Dwarka waters. All the holes have chisel marks. This has been perfectly cut. Partially buried under sediment, the exposed portion has a pinkish colour layer of marine growth and seaweed growth is also noticed. It has a smooth surface and sharp edges.
50. Grapnel anchor is made of yellow sandstone lying exposed on a sandy bottom. This anchor has been broken into two. A part consists of the upper circular hole and the other part consists of lower two rectangular holes. Lower one is broken. It has been cut perfectly. Chisel marks on the holes were also noticed. It has a smooth surface and sharp edges. A layer of pinkish marine growth and seaweed growth is also noticed.
51. Grapnel type anchor was lying partially buried in sediment. It has been cut perfectly from limestone. It has an even and rough surface with blunt edges. Sections at the lower and upper ends are rectangular. The exposed portion has pinkish marine growth with seaweed. It is one of the finest finish anchors reported from Dwarka during the present season.
52. Grapnel anchor cut out of limestone was lying exposed over rocky seabed. It has been cut perfectly. It has a smooth surface and sharp edges. As usual it has an upper circular hole and the lower two holes are rectangular. A layer of pinkish marine growth and seaweed growth is also noticed over the anchor.
53. Grapnel anchor is made out of limestone and is lying buried in sediment. It has an uneven surface and has been cut perfectly. It has an upper circular hole and the lower two holes are square. Both ends have a rectangular section. A layer of pinkish marine growth and seaweed growth is also noticed.
54. Grapnel anchor is made out of limestone and was lying exposed over a sandy seabed, surrounded by some rocky exposures. This is one of the biggest grapnel anchors found during the present season in Dwarka waters. It appears that a small upper portion has been broken as no upper circular hole is present. The section of the upper end is square and the lower end is roundish. A layer of pinkish marine growth and seaweed growth is also observed.
55. Grapnel anchor cut out of limestone was found partially buried under sediment. The lower portion of the anchor has been broken. The section of the lower end is rectangular while the upper side is roundish. It has a smooth surface and blunt edges. The exposed portion is covered with a thin layer of pinkish marine growth and seaweed growth is also observed. A few natural boulders are lying nearby.
56. Grapnel anchor cut out of yellowish sandstone and lying partially buried in sand. This is the biggest anchor found in Dwarka waters. It has been cut nicely and has a fine finish. At a few places the surface is uneven but is smooth and sharp edges. A layer of pinkish marine growth and seaweed growth is noticed over the anchor.

57. Grapnel anchor was lying partially exposed on the seabed. It has been cut out of limestone. Its upper portion has been broken. Lower two rectangular holes are present. It has a smooth surface and blunt edges. A layer of pinkish marine growth and seaweed is noticed. Sections at the lower and upper ends are rectangular.
58. Grapnel anchor made out of basalt was lying exposed on rocky seabed. It has been perfectly chiseled. Chisel marks are also noticed. It has a smooth surface and blunt edges. Sections at both ends are circular. A layer of pinkish marine growth and seaweed is noticed.
59. Grapnel anchor cut out of limestone is lying exposed on rocky seabed. Perhaps the upper portion has been broken. It has a smooth surface and blunt edges. The upper portion is roundish. Sections at both ends are circular. A layer of pinkish marine growth and seaweed is noticed.
60. Grapnel anchor made of basalt rock was lying exposed on rocky seabed. The anchor is made of a perfectly round long basalt block. Both ends of the anchor have circular sections. It has a smooth surface and sharp edges. This is the single example of this kind. It has a layer of pinkish marine growth and seaweed. As usual it has an upper circular hole and two square holes in the lower side.
61. Grapnel anchor made out of limestone was lying partially buried in sediment. Perhaps the lower portion of the anchor has been broken and only upper hole with square shape is surviving. It has a rough surface and blunt edges. The upper portion narrows down to one point while the lower end section is square/rectangular. Pinkish marine growth with seaweed is also observed over the anchor.
62. Grapnel anchor made out of limestone was lying exposed over rocky seabed just down the ridge area. It has only lower two holes while the upper hole is either broken or was not made at all. The lower hole has been widened perhaps due to continuous use or erosion. Both ends of anchor have roundish rectangular sections. It has a smooth surface and sharp edges. A layer of greyish marine growth is also observed.
63. Grapnel anchor made out of limestone was lying partially buried in sediment. Its upper portion has been broken as no upper circular hole is noticed. Section of the upper end is rectangular while the lower one is roundish or rectangular. A layer of marine growth is also noticed.

TABLE 2. INDO-ARAB ANCHORS FROM DWARKA
(DIMENSIONS IN CENTIMETRE)

Sl. no.	Year & Location	Raw material	Length	Width	Thick-ness	Dia of upper hole	Lower holes	Estimated Weight in kg
1	Before 1997	Sandstone	138	30	25	10	8 x 11, 8 x 8	
2	-do-	Limestone	120	29	26	8	13 x 16, 14 x 14	
3	-do-	-do-	117	22	25	10	10 x 11, 10 x 11	
4	-do-	-do-	92	29	26		20 x 15, 9 x 13	
5	-do-	-do-	120	26	27	- 7	10 x 10, 10 x 12	
6	-do-	Sandstone	110	27	22	12	11 x 12, 9 x 13	
7	-do-	Limestone	66	23	30		13 x 10, 10 x 10	

8	-do-	-do-	116	25	21	10	13 x 10, 11 x 10	
9	-do-	-do-	82	22	22	8	8, 8	
10	-do-	-do-	110	30	26	9	9 x 10, 13 x 10	
11	1998 IZ	-do-	85	36	24		12 x 14, 12 x 14	165
12	1998 IZ	-do-	230	40	28	15	13 x 13, 15 x 14	586
13	1998 IZ	-do-	176	38	21	8	12 x 12, 12 x 12	268
14	1998 UW	-do-	110	30	30	10	15 x 11, 15 x 10	147
15	1998 UW	-do-	116	27	18	11	15 x 15	104
16	1998 UW	-do-	108	25	20	14	14 x 14, 14 x 14	91
17	1999 B-8 UW	-do-	145	30	30	8	12 x 12, 12 x 12	216
18	1999 B-11 UW	-do-	128	28		5	12 x 12, 12 x 12	181
19	1999 B-10 UW	Basalt	142	28	27	5	11 x 11, 11 x 11	232
20	1999 B-12 UW	-do-	105	27	28	10	10 x 10, 10 x 11	199
21	1999 B-14 UW	Limestone	125	34	25	10	10 x 11, 10 x 11	206
22	1999 B-14 UW	-do-	130	35	35		14 x 14, 12 x 12	209
23	1999 B-14 UW	Limestone	188	38	30	10	10 x 10, 10 x 12	326
24	1999 B-14 UW	-do-	55	17	20	6		41
25	1999 B-16 UW	-do-	65	30	30	7		113
26	1999 B-16 UW	-do-	100	30	20	-	16 x 16	118
27	1999 B-18 UW	-do-	185	40	40	15	17 x 17, 17 x 17	443
28	1999 B-18 UW	-do-	170	35	35	5	10 x 15, 17 x 17	328
29	1999 B-18 UW	-do-	110	38	35		12 x 12, 15 x 16	310
30	1999 B-12 UW	Basalt	90	25	30		12 x 12, 12 x 12	126
31	1999 B-21 UW	Limestone	155	31	31		12 x 12, 12 x 12	20
32	2000 B-13 UW	-do-	135	25	20	10	14 x 15, 14 x 15	176
33	2000 B-13 UW	-do-	104	22		9	15 x 15	118
34	2000 B-17 UW	-do-	120	30	17		15 x 15	164
35	2000 B-17 UW	-do-	155	36	45	6	12 x 12, 15 x 13	263
36	2000 B-15 UW	-do-	147	38	25		16 x 16, 15 x 15	202
37	2000 B-14 UW	Sandstone	180	33	30	5	8 x 10, 8 x 8	264
38	2000 B-19 UW	Limestone	175	35	30	10	16 x 20, 17 x 20	188
39	2000 B-20 UW	-do-	140	33	30	10	12 x 12, 12 x 14	210
40	2000 B-21 UW	-do-	100	25	20	6	8 x 8, 8 x 8	100
41	2000 B-18 UW	Sandstone	135	28	25	6	8 x 8, 8 x 8	198

42	2000 B-12 UW	Limestone	177	35	25	10	10 x 10, 10 x 12	299
43	2000 B-9 UW	-do-	105	35	30	11	12 x 13, 12 x 13	184
44	2000 B-10 UW	-do-	108	28	23		10 x 12, 10 x 12	122
45	2000 B-8 UW	-do-	143	25	28	7	12 x 13, 10 x 10	192
46	2000 B-24 UW	-do-	95	36	28		14 x 12, 14 x 16	182
47	2001 B-21 UW	-do-	125	40	30	15	12 x 12, 12 x 12	180
48	2001 B-17 UW	-do-	129	25	27	11	12 x 12, 11 x 12	82
49	2001 B-12 UW	-do-	191	44	41	13	13 x 13, 15 x 15	562
50	2001 B-31 UW	-do-	170	30	23	11	8 x 11, 11 x 12	216
51	2001 B-38 UW	-do-	155	34	27	6	10 x 11, 7 x 7	219
52	2001 B-18 UW	-do-	130	30	38	10	9 x 11, 9 x 11	229
53	2001 B-45 UW	-do-	132	30	26	10	13 x 10, 11 x 10	
54	2001 B-42 UW	-do-	80	26	22	6	—	
55	2001 B-41 UW	Limestone	102	25	24	6	12 x 10, 11 x 10	371
56	2001 B-40 UW	-do-	238	40	30	14	12 x 12, 12 x 12	697
57	2001 B-12 UW	-do-	100	20	18		10 x 9, 10 x 11	103
58	2001 B-19 UW	Basalt	135	30	30	5	12 x 10, 11 x 9	184
59	2001 B-22 UW	Limestone	85	25	20		11 x 8, 8 x 11	65
60	2001 B-25 UW	Basalt	114	30	37	7	10 x 10, 10 x 10	12
61	2001 B-47 UW	Limestone	125	42	28		11 x 11	283
62	2001 B-33 UW	-do-	100	29	18		17 x 20, 13 x 12	111
63	2001 B-7A UW	-do-	76	32	28		13 x 12, 14 x 10	110

* Actually valued
Value of basalt per m3 = 2900 kg, Value of sandstone per m3 = 2300 kg
Value of limestone m3 = 2500 kg

4.5 RING STONE ANCHORS

A large number of ring stone anchors have been found from Dwarka and Somnath. The anchor is characterized with spheroid shape, a wide hole in the middle and they have been reported mainly from the Gujarat coast (Rao, 1999, Gaur *et al.*, 2002) and from Oman (Vosmer, 1999, personal communication). Subsequently, many more ring stones have been found from Dwarka and Somnath (Technical Report, 2001) waters. The presence of these ring stones in great numbers in Somnath and Dwarka provided an opportunity to discuss further on their uses, origin and date, which have been dealt with in detail here.

Offshore explorations at Dwarka and Somnath brought to light a large number of different types of stone anchors such as composite, grapnel and ring stone types besides a few stone structures in Dwarka waters, out of which ring stones from Dwarka are 24 and from Somnath are 35. These ring stones have been collected during various field trips to respective sites. The basic characteristics of ring stone anchors are circular shape with an axial hole. Often the base

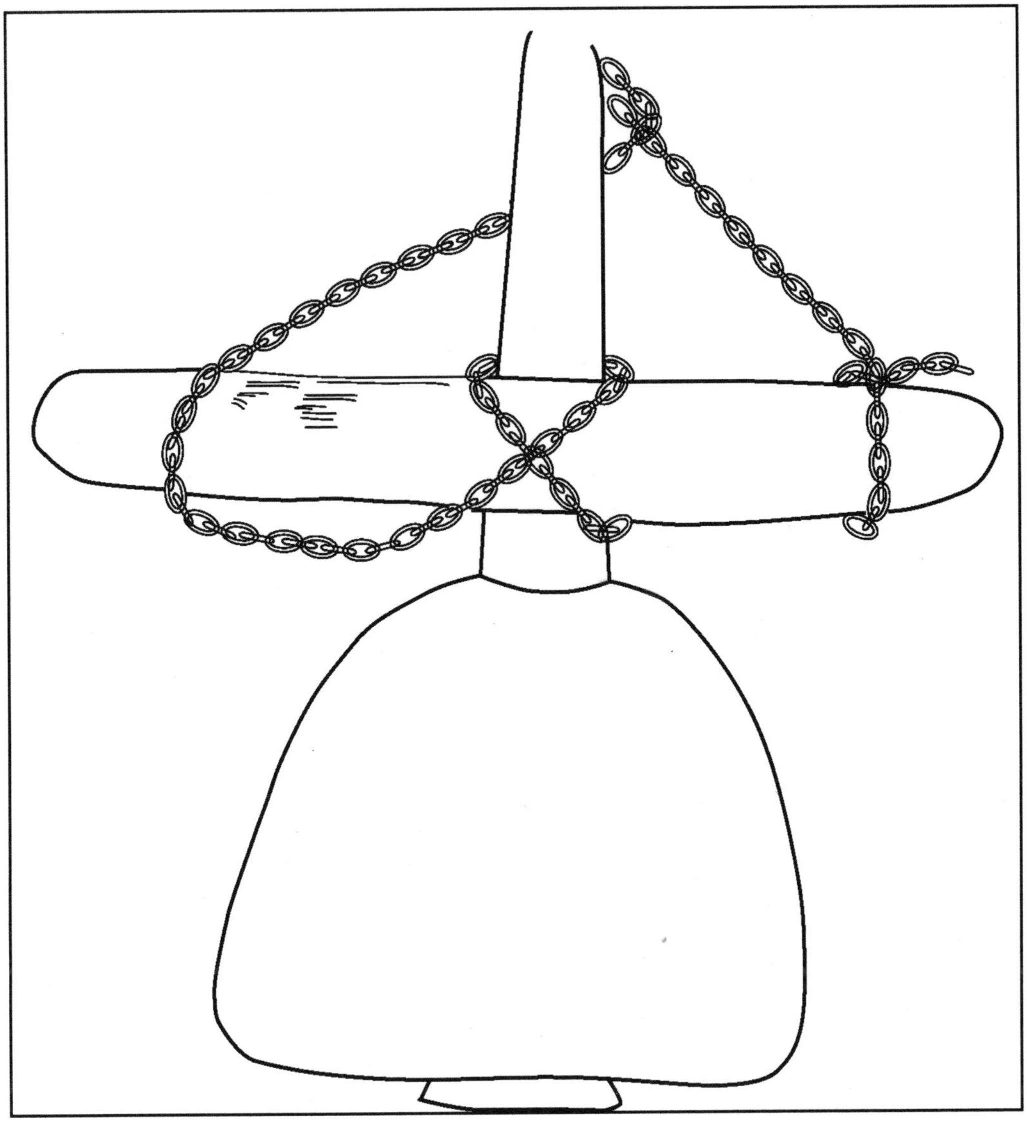

Fig. 18. Hypothetical reconstruction of ring stone as an anchor.

of the ring stone is flat and top semicircular rising to a certain height. The detailed description of the ring stones from each site is given as follows:

Offshore explorations during the last two decades have brought to prominence a number of ring stones from Dwarka area. An important observation may be mentioned here that most of the objects have been found between 8 to 16 m water depth, few in shallow depths and one instance from intertidal zone. The positions of the ring stones discovered during the last two seasons have been plotted on the map. A total of 24 ring stones have been recovered from Dwarka waters (Table 1). The anchor nos. 1 to 11 were recorded before 2000 AD and remaining from 12 to 24 were recorded in the season 2001 AD.

Most of the ring stones were lying exposed on seabed; however, a few ring stone nos. 2, 5, 10, 16 and 22 were lying partially buried in sediment. The exposed portion of the anchors till a depth of 8 m was covered with marine growth such as seaweed while beyond this depth ring stones are covered with a thin layer of greyish marine growth. They are lying in vertical position, which is the normal position of anchor falling on the seabed and few ring stones were tilted like nos. 14, 17 and 22. A few ring stones have also evidence of chisel marks on the surface, in the hole and on the flat bottom side. On the basis of material and shapes, the ring stones from Dwarka can be broadly divided into three groups.

4.5.1 Group I

Ring stone nos. 2, 4, 5, 9 and 12 fall in this group. The raw material used for this category is basalt and they are heavy in weight. The basic features of this category include less height, flat bottom, semi-curved top with an axial hole and the diameter of the hole is same on both ends. Chisel marks have been noticed on the surface and in the hole. The average estimated weight of the ring stones is 147 kg. Among them, the heaviest one (no. 4) weighs 245 kg and the smallest (no. 12) 80 kg.

4.5.2 Group II

Ring stone nos. 3, 7, 8, 10, 13-17, 19-21, 23 and 24 fall in this category. The majority of ring stones of this category are made out of limestone and sandstone. The shape of these ring stones are same as of the first category; however, often they have blunt edges and a widened axial hole. A few anchors have a smooth surface, but others have very rough surface also. Sometimes chisel marks are also noticed on some of the anchors. The average weight of this group is 98 kg. However, the heaviest anchor (no. 8) weighs 205 kg and the smallest (no. 14) 64 kg.

4.5.3 Group III

Ring stone nos. 1, 6, 11, 18 and 22 fall in this last category. The ring stones are made out of porous limestone. As usual they are circular in shape of comparatively less height and have equal diameter on the upper and lower side ends. Often axial holes are wide and irregular. They have blunt edges and a rough surface. The average estimated weight of these ring stones is 66 kg. The heaviest anchor weighs 109 kg and the lightest anchor 20 kg.

4.5.4 Discussion

The above description of ring stones suggests that they are similar to each other in shape; however, the size and raw material varied at both the places. The possible uses, date and probable origin of these ring stones is discussed here. Before a detailed discussion, a few facts about these objects must be considered.

1. These are man-made objects as several ring stones have the remains of chisel marks in the axial hole and on the surface, and have not been found in association with any habitational site in Somnath as well as Dwarka.
2. Their association with composite and grapnel stone anchors indicate that these stones were used for navigational purposes.
3. These ring stones were often found in rocky seabed.
4. These anchors have been found in large numbers in Saurashtra coast, which seems to be a place of the origin of these ring stones.

4.5.4.1 Ring Stones from Other Places

A number of similar type of objects have been reported from Mohenjo-Daro (Marshall, 1973 and Mackay, 1976) and Harappa (Vats, 1975) where they were described as limestone rings, which have very smooth surface and were used for either making pillars or for some ritual

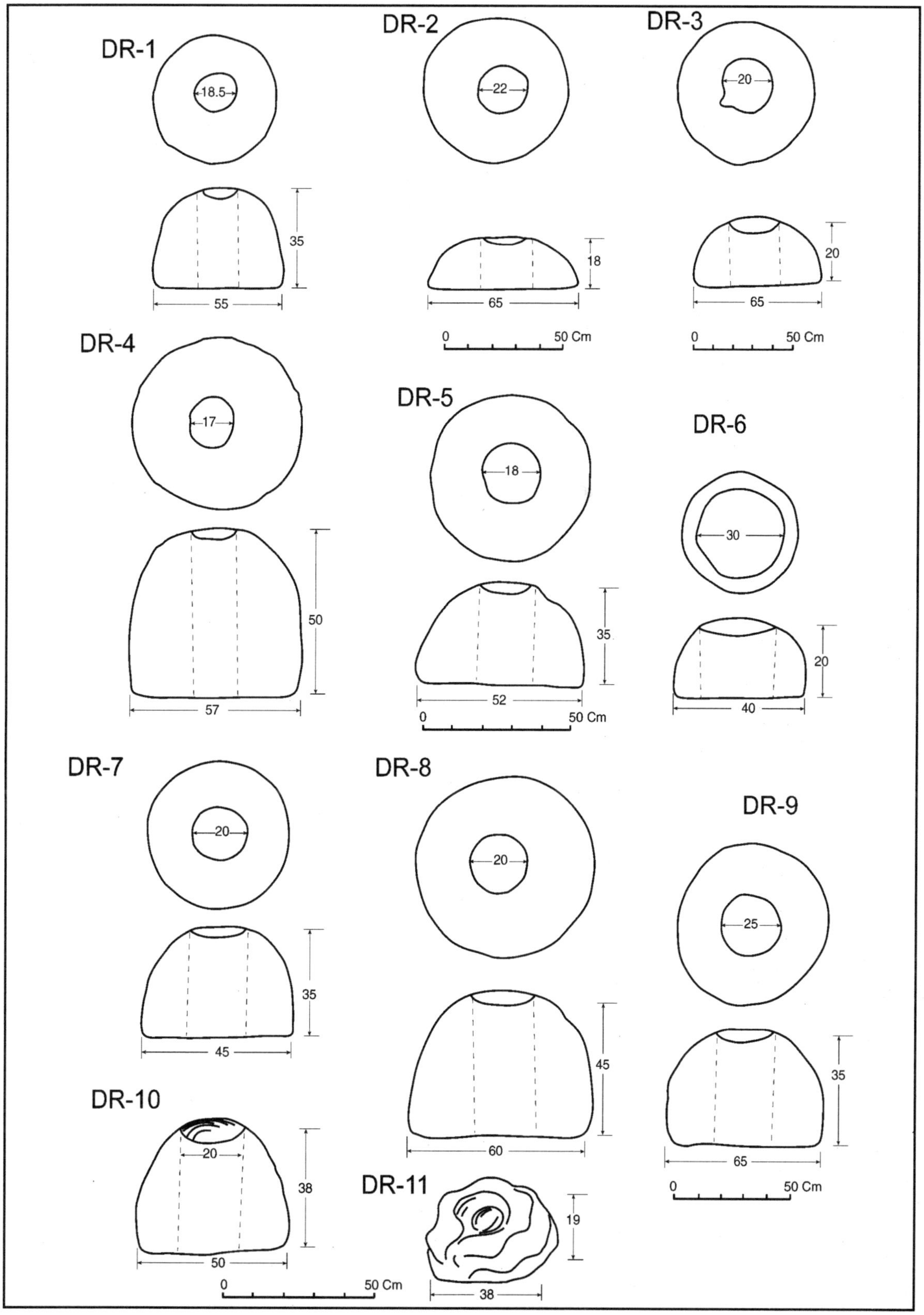

Fig. 19. Ring stone anchors No. 1 to 11 from Dwarka waters.

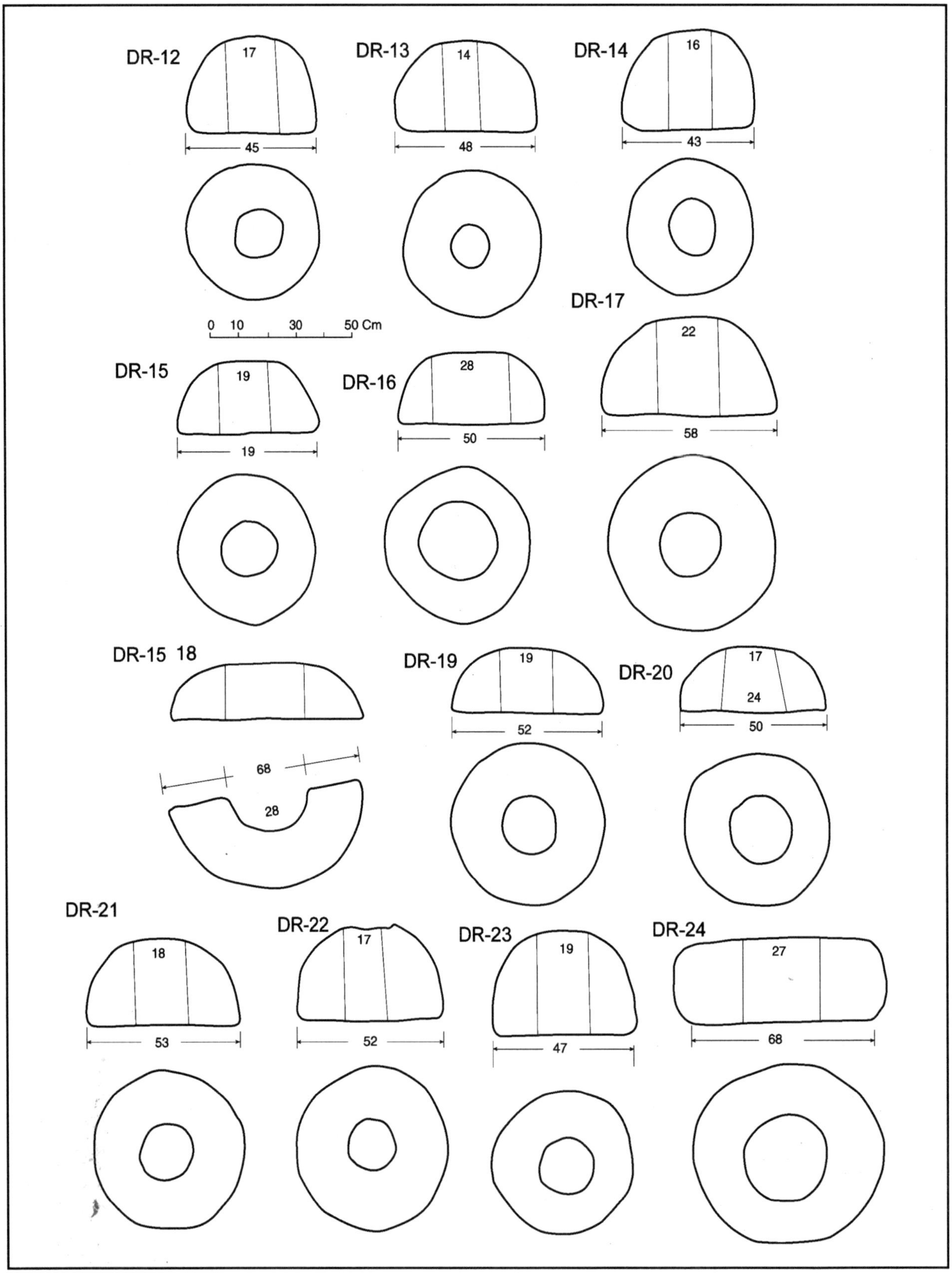

Fig. 20. Ring stone anchors No. 12 to 24 from Dwarka waters.

purposes. Marshall (1973) mentioned that the most typical of them have their upper and lower surface undulating, while in others, the lower surface is flat, and the top takes a quatrefoil form. He believes that these objects were used either for ritual or architectural purposes. About the ring stones from Mohenjo-Daro, Mackay (1976) says:

> the height of the ring and the diameter of the central hole increases more or less progressively with the diameter of the ring. The ring stones of Mohenjo-Daro also have dowel holes. On the three the rings stones there were two holes in the top and two in the base, averaging ½ inch in dia. 1 inch deep and set equal distances from the edges of the central hole.

However, this phenomenon has not been observed in the case of Dwarka and Somnath. A similar type of stone object is reported from Sri Lanka (Souter, 1998) with square axial hole and has been described as ring stone anchor. There are examples from Maldives waters where similar type of objects have been recovered and described as ring stone anchors (Vosmer, 1999, personal communication). Six ring stones have been discovered off Oman and it is suggested that these objects were used to anchor large fishing nets or traps until the 1950s. Given here are different views have been suggested on the uses of these ring stones.

4.5.4.2 Various Views on the Uses of Ring Stones

Rao (1999) has suggested the use of these objects as doorjamb as they were found along with the stone structures off Dwarka. However, this observation is not true as the objects have been found along with other stone anchors where stone structures were not noticed. The recent explorations off Dwarka have also indicated that the stone structures may be the remains of an ancient jetty or harbour (Technical Report, 2000).

Ethno-archaeological explorations in Oman suggested that these ring stones could have been used as anchors for fishnet. The size of the stone used depended on what type of fish was being caught (Vosmer, 1999, personal communication). The local fishermen of Oman suggest that 5 to 7 stones were used at a time and they were not more than 40 kg in weight. However, the size and the weight suggested by the fishermen do not match with the present available ring stones in India as well as in Oman. The weight of ring stones in India varies from 392 kg to 16 kg and by no method any fisherman would be using ring stones weighing so heavy for fishnets. Another point may be mentioned here that if these stones are used as per suggestions made by the fishermen of Oman, it should be only for very shallow waters whereas these objects have also been found at 15 m water depth off Dwarka and Somnath. Out of 59 ring stones, 14 are below 50 kg, 22 are more than 50 kg, 17 are more than 100 kg, and 6 more than 200 kg. Therefore, these ring stones may not be suitable as a weight for fishnet. Hence, the possibility of their as fishnet weight may be ruled out.

There is a strong possibility about the use of these ring stones as anchor for boats (Gaur *et al.*, 2002). Vosmer (personal communication, 1999) informed us that similar types of ring stones have been discovered in Maldives waters and also suggested the use of these as anchoring boats in coral reef bottom. It seems that these ring stones were used as anchor of boats in the disturbed rocky bottom. Two objects off Oman, 35 off Somnath and 24 off Dwarka have been noticed in the disturbed rocky bottom. There may be apprehension about the use of ring stones, especially those which have less weight (less than 50 kg) and might be a fishnet weight; however, the size of anchors always depend on the size of boats. Even today a stone weighing less than 15 kg is used by small out-trigger boats on Konkan coast and other coasts of India. Therefore, the less weight of a few ring stones does not indicate that their function would be different from the anchors. It is possible that one boat might be carrying all types of stone anchors and they were using as per the requirement of sea bed as all of them have been found together in Dwarka and Somnath waters. If it was sandy, certainly composite anchors were more effective, and in rocky areas where there was not much grip, heavy prismatic anchors were in use and the most disturbed area ring stones were very much effective.

With all probability, it may be said that ring stones would have been used as anchors for boats. The shape of the ring stone is spheroid and the flat part would serve as lower part while the curved portion would be the upper side. A wooden log would be inserted in the axial hole with thickened portion on the lower side. Often the lower edge of the axial hole is blunt, which may be due to the thickened part of wood's interaction with the lower edge while the upper side wooden log will not affect the edges, and therefore, the edge of upper side is sharp. The central wooden log will be further attached to a cross wood which will be tied with a rope (Fig. 18), so whenever the ring stone is thrown in water, it would fall with its flat base down, semi-curved portion above and the rope of boat will be on the upper side and would not entangle with any rock outcrop. The findings also suggest that most of the ring stones have been found in the same position.

4.5.4.3 Probable Date

Dating of these ring stones would be speculative as they have not been found associated with any datable archaeological context. Scientific dating may not be possible in absence of wood remains in the hole of the ring stones. However, they have been found with composite and grapnel stone anchors. Therefore, a date ranging between the 8th and the 17th century AD will be more favourable as suggested for other stone anchors (Sundaresh *et al.*, 1999 and Gaur *et al.*, 2001). A reference has been made by Schoff (1974:155) about a visit to Oman by Gemelli Carreri, in 1693-69, and he mentioned that, "for anchor, they have a large stone with a hole". This could be a ring stone and if the reference is for this type of anchor then this will a lone authentic record of the 17th century about the ring stones. However, these must have continued since very early times in Dwarka and Somnath waters. The majority of these objects are made out of locally available rock and a few are of hard rock (may be basalt) indicating that these were made indigenously.

4.5.4.4 Origin of Ring Stones

The largest number of ring stones have been reported from Somnath waters, followed by Dwarka. As mentioned above, a few instances have come from Maladives and Oman and absent in Red Sea and Mediterranean Sea (as various nautical archaeological publications have not published these types of stones). Perhaps these must have been invented by the Indian sailors with gaining experience of seabed specially in Gujarat coast where submerged reef and uneven seabed poses the problem for lifting heavy grapnel and composite type of anchors. Another important point may be mentioned here that these anchors need a very thick wooden stock, which will not be easily lost on the seabed as compared to other types of anchors where a thin wooden log was inserted. The present data suggest that ring stone anchors are the speciality of Saurashtra coast of Gujarat, of Medieval Period. Therefore, it is proposed, a new nomenclature as 'Saurashtra stone anchors' for these stone anchors. The sailors of this coast must have visited Oman and Maldive from where similar types of ring stones have been reported.

4.5.4.5 Description of Ring Stones (Figs. 19-20; Pls. 36-38)

1. The anchor was made of sedimentary fine-grained rock and three layers can be observed. The upper and lower ones appear to be reddish while the middle one is yellowish in colour. It has a tapered apex with an axial hole. The exterior of the object is irregular, probably as a result of due to erosion.
2. This anchor was lying in 7 m water depth. It is made out of basalt and has a flat bottom. Chisel marks were also noticed. It has a smooth surface and fine edges. This ring stone has a low height and higher diameter as compared to other ring stones.
3. The ring stone anchor was partly buried in sediment. It has a flat base with an axial hole. This is cut out of basalt. It has fine edges and a rough surface. Chisel marks are also noticed.
4. This ring stone anchor was lying in 10 m depth. It is made out of limestone. It has a smooth surface and sharp edges, flat bottom, tapering upper portion, and an axial hole with 15 cm diameter. It was found lying in a channel.
5. The ring stone was made out of sandstone. It has very fine edges. The bottom is flat and show tapering on the upper side with an axial hole. Chisel marks in the hole are also noticed.
6. This ring stone was noticed near the previous one. Made of limestone, it has a very rough and pitted surface. The shape is also irregular; however, a wide axial hole is present.
7. Another ring stone was lying partially buried in sediment. It was exposed by airlift operation. This is made of basalt, and has a smooth surface and sharp edges. Chisel marks are also noticed. It has a flat bottom and a tapering upper portion. An axial hole is also noticed.
8. This anchor was lying in the intertidal zone of Dwarka. It is made out of conglomerate rock of reddish colour. It has a rough surface and chisel marks in axial hole are also noticed. This is the biggest ring stone anchor found from Dwarka area.
9. The ring stone was lying in shallow waters near the stone structures opposite river Gomati. It is made of hard rock (basalt?). It has a smooth surface and is partially buried in the sediment. As usual it has a circular shape with tapering upper portion and an axial hole.
10. The anchor was found buried in a channel at a water depth of 15 m. It has a circular axial hole and the bottom is flat while the upper portion is tapering. A small hole at the upper side, which opens to a bigger hole, is also noticed. It is made of limestone and has a very rough and porous surface.
11. The anchor was noticed in shallow waters. It is irregular in shape with a rough surface. It is made of locally available sandstone. A wide middle hole and a small side hole is also noticed.
12. The anchor was lying tilted and partially buried in sediment. This has been perfectly cut out of basalt. Chisel marks have been noticed in the hole as well as on the outer surface. As usual it tapers on the upper side. It has a smooth surface and the edge of the upper end is sharp while one on the lower end has been blunted. The exposed portion has a layer of pinkish marine and seaweed growth.
13. The ring stone anchor was lying exposed on rocky seabed. It is made out of limestone. The anchor has a very fine finish. The bottom of the object is flat and smooth and the outer surface is also smooth with sharp edges. A thin layer of pinkish marine and seaweed growth is also noticed over it.

14. The ring stone anchor made out of yellow sandstone was lying partially buried in sediment. It has been cut perfectly and tapers on the upper side. Chisel marks are also noticed. The surface is uneven but smooth. The anchor was lying upside down. A layer of pinkish marine and seaweed growth is observed.
15. The ring stone anchor was lying exposed over sandy seabed. It has been cut out of limestone and chisel marks on the hole are noticed. The anchor has a rough surface and blunt edges. It has a wide middle hole, flat bottom and tapered upper side. A layer of pinkish marine and seaweed growth was also noticed.
16. The ring stone anchor made out of limestone, was lying exposed over the rocky seabed. On one side of the anchor the height has been reduced perhaps due to erosion or extensive use. The outer surface of the object is irregular but has smooth surface and sharp edges. The hole has been widened. A layer of pinkish marine and seaweed growth has been noticed.
17. The ring stone anchor made out of limestone was lying exposed over the sandy bottom. It has a very irregular shape. The anchor has an uneven and rough surface. On the lower side pinkish and upper side whitish marine growth has been observed. The lower side hole has been widened.
18. A half-broken ring stone anchor was lying exposed on the rocky seabed. The anchor is made out of grey sandstone. The anchor has an irregular shape and its height is less. It has a smooth surface and blunt edges. A layer of pinkish marine and growth of seaweed over the object have been noticed.
19. The anchor was lying exposed over the rocky seabed. It has been cut perfectly from a limestone block. The anchor has a rough and porous surface with blunt edges. A layer of pinkish marine and seaweed growth is noticed. Both ends of the hole have been widened.
20. The ring stone anchor made out of limestone was found exposed over the sandy seabed. Its exterior is irregular and uneven. The anchor has a rough and porous surface and blunt edges. The lower end of the hole has been widened. A layer of pinkish marine and seaweed growth is noticed.
21. The limestone anchor was lying exposed over the sandy bottom. It has a very irregular shape. The anchor has an uneven and porous surface. The axial hole is also irregular. The hole at the bottom has been widened. A layer of pinkish marine and seaweed growth is also noticed.
22. The ring stone anchor cut out of yellow sandstone, lay buried in sediment. It has a flat bottom but the height is low and does not taper on the upper side as other ring stones. It has an uneven and smooth surface. It has been cut very nicely. The lower end of the hole has been widened. A layer of pinkish marine growth and seaweed is also noticed.
23. The ring stone anchor is made out of grey sandstone and was lying exposed over the rocky bed. It has a rough surface and blunt edges. Chisel marks in the hole were also noticed. As usual it tapers towards the upper side. A thin layer of pinkish marine growth is also noticed.
24. The stone anchor was lying exposed on the rocky seabed. It is made out of limestone. It has a porous surface and blunt edges. The anchor is flat on both sides and does not taper as other ring stones. A layer of pinkish marine and seaweed growth is also noticed.

TABLE NO. 3. RING STONES FROM DWARKA WATERS
(DIMENSIONS IN CENTIMETRE)

Sl. no.	Year of find	Location	Raw material	Total diameter (cm)	Diameter of hole (cm)	Height (cm)	Weight in kg (estimated)
1	1988	-	Sandstone	55	18.5	31	109
2	1992	SS-12	Basalt	65	22	18	111
3	1992	SS-28	Limestone	65	20	20	134
4	1998	Zone 2	Basalt	57	17	50	245
5	1998	Zone 2	Basalt	52	18	35	13
6	1998	Zone 2	Limestone	40	30	20	20
7	1997	Zone 5	Limestone	45	20	35	94
8	1997	Intertidal zone	Conglomerate rock	60	20	45	205
9	1990	E-10	Limestone	65	25	35	215
10	2000	B-11	Limestone	50	25	38	136
11	2000	B-23	Sandstone	38	11	19	33
12	2001	Buoy 44	Basalt	45	17	28	80
13	2001	Buoy 37	Limestone	48	14	27	85
14	2001	Buoy 36	Sandstone	43	16	31	64
15	2001	Buoy 46	Limestone	50	19	21	67
16	2001	Buoy 28	Limestone	50	28	30	77
17	2001	Buoy 29	Limestone	58	22	28	120
18	2001	Buoy 11	Grey sandstone	68	28	18	90
19	2001	Buoy 34	Limestone	52	19	19	66
20	2001	Buoy 30	Limestone	50	24	25	72
21	2001	Buoy 23	Limestone	53	18	21	78
22	2001	Buoy 13	Yellow sandstone	52	17	25	79
23	2001	Buoy 24	Sandstone	47	19	28	68
24	2001	Buoy 26	-do-	68	27	20	102

5
Aramda and Positra in Okhamandal

5.1 ARAMDA

The village Aramda is situated on a high tableland about 7 km south of Okha port and 23 km north of Dwarka. A creek opens to the sea in the northeastern side of the village. During high tide, a vast area experiences floods. This is also a fishing jetty and served as a repairing yard. For some time the village of Aramda had been a power centre of the Okhamandal region, as is attested by the presence of a medieval fort in the middle of the village.

Onshore and intertidal zone explorations were carried out to find out any possible signature of an ancient port or harbour in the region. An old jetty, that is still serving as repairing yard is located on the northeastern side of the village. Other important findings include anchors, a dockyard (Pl. 43) and boat motifs that are described below.

5.1.1 Anchors

Two stone anchors were found to the north of the present jetty. Their brief description is given below.

1. Upper portion of the anchor is broken and only one rectangular hole at the lower side of the anchor is noticed. It is made of locally available limestone rock.
2. A triangular anchor with three holes was noticed during low tide with a circular hole at the top and two square holes on the lower side (Pl. 41). The anchor now serves as a mooring stone for fishing vessels.

An old dockyard was noticed to the southern side of the present jetty. It had a stone wall covered with mud. This is rectangular in shape and gets submerged during high tide. The length was about 125 m with a width of about 30 m. The present depth of the dock is about 1 m. There is an entrance from the southeast which is silted up and the side walls have collapsed and presently it is not in use.

5.1.2 Boat Motifs

Several hero stones were noticed at the extreme west of the village Aramda (Pls. 39 & 40). Three of them had boat motifs. Locally this place is called Jhapa. Hero stones with boat motifs are known as Kharwa's memorial. The Kharwas were the early navigators and perhaps these stones were erected in their memory. One of them is dated to *Vikram Samvat* 1852 (AD 1796) and the boat motif is different than the other two hero stones. The motif has several sails and is similar to those of long sea voyage vessels. The other appears to be of a later period. These boat motifs indicate the importance of those persons involved in maritime trade and commerce in this region.

The discovery of stone anchors and boat motifs indicates that this was an important maritime centre. The site is well protected from high waves of the open sea and is an ideal sheltered harbour. However, archaeological evidences so far from this village do not indicate its age earlier than the Medieval Period.

5.2 POSITRA

Positra is famous for pearl fishing in the Okhamandal region. Onshore exploration revealed a few hero stones to the extreme west of the village Positra and this place is called Jhapa. One of them had a boat motif (Pl. 42). Local information suggests that these memorial stones were erected in the memory of seven persons of this village who died due to boat-capsizing. These memorial stones are engraved with boat motifs and horse riders. The boat motifs on the hero stones indicate the importance of those persons involved in maritime activity in this region.

A late medieval site was noticed at Dablo about 4 km on the eastern side of the village Positra. A large number of the late medieval pottery were collected.

6
Conclusions

The concentration of the dressed stone blocks is observed only at one location (Zone 1) opposite the mouth of the Gomati creek. Apparently, the construction of buildings was limited to this area and the port town did not cover a large area. Anchors found beyond these structures suggest that boats were anchored away from these construction sites. A stone block with Gujarati inscription indicates a later date for the stone structures.

Underwater explorations carried out in a very large area off Dwarka have revealed the remains of an ancient jetty and a large number of stone anchors and these evidences suggest that Dwarka was one of the most active port during the Historical and Medieval periods. Underwater investigation has not revealed any protohistoric settlement (3500 years BP) and the submergence of ancient Dwarka is yet to be ascertained. The offshore area has been thoroughly scanned up to 25 m water depth to observe remains of any submergence ancient settlement but none were found. A study of coastal Harappan sites of Gujarat (Gaur and Vora, 1999) and sea level (Merh, 1992) suggest that during the Harappan Period, the sea level was higher than the present (between 2-6 m). Therefore, sea level history also does not support the hypothesis of submergence of any archaeological settlement dating back to the mid-2nd millennium BC, though there is evidence on the minor sea level fluctuation during the early centuries of the Christian era along the Bet Dwarka coast (Gaur *et al.*, 2007).

The present work has yielded some different evidences compared with the earlier researches and whatever remains that have been noticed during the present exploration, they do not suggest that submerged structures of Dwarka belong to some habitation site; they rather appear to be the remains of a jetty. Circular and semicircular structures are remains of the pillars, over which the super structure of the jetty rests. Sankalia (1966) in the report of "Dwarka Excavation" mentioned the existence of a busy sheltered harbour close to the Dwarkadhish temple. A painting dating back to the mid-19th century suggests that large vessels used to anchor near present Gomati Ghat. A study of the surrounding archaeological sites indicates that the submerged structure may not be as old as suggested earlier and datable to 1500 BC. However, the date of these structures may be still a matter of debate. A large number of stone anchors from Dwarka waters indicate that Dwarka was one of the most active ports in the past. Recent discoveries of stone anchors from all over the coasts of the Indian Ocean suggest that Dwarka anchors may be dated between the Early Historic Period and the Medieval Period.

Onshore and intertidal zone explorations in Okhamandal suggest that this was basically occupied due to richness of marine resources, which may have been exploited considerably. The excavator of Nageshwar argues that the Harappans at Nageshwar were influenced by the Sindhi Harappan and termed the site 'Sindhi Harappan' and they were not having contact with the Saurashtra Harappans, whereas settlement at Bet Dwarka was largely influenced by the Saurashtra Harappans as indicated by the potsherds. Okhamandal area was continuously occupied at different places. The most of the present villages came into existence during the Medieval Period.

Underwater explorations in Somnath waters brought to light a large number of ring stone anchors and these findings are very similar to those found at Dwarka. Therefore, the period of these artefacts may be the same as of Dwarka. Somnath and Dwarka witnessed occurrence of the largest number of ring stone anchors and it appears that these anchors were invented by the Saurashtra mariners; therefore, a new nomenclature is proposed as "Saurashtra stone anchors".

At last, it may be said that while searching an ancient celebrated city one should not be confined to or pinpoint a particular place. In case of Dwarka, one may consider the entire Okhamandal area as one geographical entity, which is separated by the Okha Rann. The oldest remains are lying at Nageswar dating back to the Mature Harappan phase whereas Bet Dwarka represents the late phase of the Indus Civilization. Rest of the villages of Okhamandal area came into existence either during the Historical or the Medieval Period. Interestingly, most of the villages of Okhamandal region have some religious significance.

With the above note, one can place Dwarka of the Mahabharata period at any of the above-mentioned places. So far destruction of Dwarka by sea is concerned, it can be logically argued with the occurrence of recent natural phenomena on the eastern coast of India called 'tsunami' which can destroy any village of Okhamandal area. However, the remains of such destruction need to be looked far.

Section II
SOMNATH

1
Introduction

Like Dwarka, Prabhasa-Somnath is not only a holy place but also a site of the historical importance. The archaeological excavations on land at Prabhasa by the Baroda University and Deccan College, Pune, have yielded evidence of a township dated to 2000 BC. The temple of Somnath is dedicated to Lord Shiva and the *linga* is counted among the 12 *Jyotirlingas* of India mentioned in the various *Puranas*. Prabhasa is mentioned in the *Mahabharata* as Prabhasatirtha (118.15, 119.1-3). The *Bhagavata Purana* says that at this sacred place the river Saraswati flows towards the west and the town is situated on the seashore (X.45.38; X.78.18). The same text also says that Lord Krishna undertook a sea voyage from Bhrugukachcha to Prabhasa. Nasik and Karle cave inscriptions of Nahapana (AD 119-124) also refer to Prabhasa (*Epigraphia Indica*, 1981). Prabhasa, which seems to have extended up to Somnath, was a flourishing port during the Protohistoric times. After a short period of decline, the port again became prominent during the Early Historical and Rajput periods. Al-Beruni mentions that China and Zanzibar had trade contacts with Prabhasa and the main exports were cotton, textiles and leather goods to Arabia and imports were corals, pearls, ivory tusk, diamond and clover into Gujarat. Thus, the ports of Gujarat including Somnath flourished due to extensive overseas trade with the Southeast Asia, Ceylon and the Middle East countries. The results of earlier onshore and offshore excavations are described below.

1.1 COASTAL GEOMORPHOLOGY

The Prabhasa-Somnath coast forms a straight line revealing a vivid manifestation of marine aeolian and fluvial processes that have resulted in a number of important geomorphic land forms, as the near-shore zone is characterized by the formation of recent alluvium deposits, sand bars, mud flats and mangrove swamps. The coastal zone is covered with beaches and littoral sand, oyster beds and sand dunes.

1.2 Land excavation

Prabhasa was excavated by the Department of Archaeology of Gujarat and M.S. University of Baroda in 1956 and 1965 and further excavations carried out by the Deccan College, Pune from 1971 to 1972 and 1975 to 1976 revealed four cultural (IAR, 1971-72, 1975-76).

1.2.1 Period I

This period is represented by a coarse Red and Grey Ware, which is associated with a Red-Slipped Ware and a Black and Red Ware. The Red and Grey pottery is represented by *Kundas* with flaring sides and a flat bottom. The Red-Slipped Ware is of fine fabric and also highly burnished. The Black and Red Ware has its exterior completely red and the interior completely black – a feature characteristic of the Black and Red Ware from Lothal (Sankalia, 1974). Besides this, there is painted Black-on-Red pottery similar to the late Harappan in Gujarat. Among the small finds, mention must be made of segmented faience beads and a blade flake industry of chalcedony (IAR, 1971-72).

1.2.2 Period II

This period is characterized by thc Prabhas Ware. The vessels in the ware are of fine fabric and are treated with a pinkish or orange slip, which, in several cases, have turned grey because of peculiar conditions of firing in the kiln. Over the surface are painted patterns in purple or dark brown. The shapes in the Prabhasa Ware mostly copied from Harappan pottery such as the dish-on-stand, big storage jar with heavy beaded rim and dish. But the most prominent type is a sub-spherical bowl with a featureless rim. The habitation remains include a house, which is rectangular on plan built of large flat slabs of meliolite limestone, which abounds in the region (IAR, 1975-76).

1.2.3 Period III

Prabhasa Ware of Period II continues in this period but Lustrous Red Ware begins to occur now. The most noteworthy feature of this period is the discovery of an extremely complex structure built of boulders of miliolite limestone set in mud mortar very close to the mouth of the river Hiranya. A close examination of the artefacts recovered from the structure indicates that it was a public building, perhaps the warehouse or market place of a port town (Rao, 1991). The warehouse can be compared with the West Asian (Tepe Yahya) warehouse (Ghosh, 1989). The most noteworthy find of this period is a steatite seal amulet, which is engraved on both sides. On one side there are seven stylized sheep, and on the other only five.

1.2.4 Period IV

This period represented by the Early Historical pottery such as the Black and Red Ware and fragments of amphora suggests trade contact with the Roman world.

1.3 Onshore exploration in 1991

Explorations carried in and around the excavated area and on the bank of the river Hiranya by the Marine Archaeology Centre of NIO yielded pottery of different period including the Late Harappan pottery and Red Polished Ware. The nick point of the Hiranya River where it used to join the sea in now buried under sand-bars.

Near-shore exploration: Before undertaking offshore exploration, the near-shore region and the intertidal zone were explored from Veraval jetty up to the sand-bars near the mouth of the river Saraswati. During ebb tide near Veraval six *lingas* came to light. A few of them are worshipped today. A shipwreck is also visible during the ebb tide in 0.5 m to 2 m water depth. Local inquiries revealed that the wreck is hardly 10-15 years old.

1.4 Offshore explorations by MAC in 1991

Offshore explorations were carried out west of the Somnath temple and opposite the river Hiranya. Diving operations were started initially in shallow waters and gradually extended towards the deeper waters up to 9 m water depth off Somnath. The exploration area has been divided into 3 zones and diving was carried out at 15 places. In Zone 1 circular stone artefacts with axial hole in the centre were found lying on the rocky patch. A few of them were partially buried in sediment. A 2-m deep channel with a prismatic stone anchor laying in it was observed. The whole area is overgrown with soft pink and grey corals. Zone 2 is situated opposite to the Somnath temple. The area is full of sand and no vegetation was noticed. In Zone 3 the prototype of corals was encountered. The purpose of diving here was to search for submerged channel of the Hiranya River, if any.

Underwater explorations at Somnath revealed triangular and prismatic stone anchors, which are similar to Dwarka type. Dressed stone blocks and massive single-holed circular stone artefacts of different shapes lie in the rock-cut channel. The large spheroid stones measuring 75 x 37 x 25 cm and 88 x 61 x 28 cm and having a hole appear to be flag-posts used on boats or on the shore. Further exploration may reveal some more archaeological remains to arrive at a conclusion whether or not a township of 1500 BC lies submerged. The artefacts so far found appear to be of a later date. If in the proposed exploration pottery or other antiquities are found, the age of the submerged objects can be determined.

2
Underwater Explorations

Offshore explorations have been carried out in a vast region in the southwest direction of Somnath temple (Fig. 21). The entire area can be divided into 3 zones. A brief description of seabed topography and archaeological finds are discussed here.

2.1 ZONE 1

The area falls under shallow water depth, on the west of the Somnath temple (Pl. 44). Water depth ranges between 5 to 7 m. The seabed consists of fine white sand and a few places have rocky formation; however, on the further western side of the site a big rocky patch appeared where a few channels were also noticed (Pls. 45-46). These rocky patches are covered with seaweed and other marine growth. Five ring stones were noticed in this area. Among them one was lying exposed over the rocky bed and the remaining four were partially buried in sediment (Pl. 47). Two were made of basalt while the other three were of limestone. All of them have a smooth flat bottom and taper towards the upper side. A wide axial hole in the centre was also noticed. Two objects have chisel marks in the axial hole as well as on the surface. One of the anchors made of limestone has been eroded significantly. The exposed portion of these objects is covered with marine growth.

2.2 ZONE 2

This area falls on the southern side of the Somnath temple, measuring about 500 x 500 m. Water depth ranges between 9 m and 13 m. The seabed consists of disturbed rocky bed. At few places, the rock bed is plain. A few channels were also noticed. A number of archaeological objects were recorded in this zone. They include 10 ring stones, 4 triangular stone anchors, 1 grapnel stone anchor and a marble statute of a goddess. All these are lying in a very vast area.

Among 10 ring stones, one is made of haematite; it was partially buried and lay upside down. Chisel marks in the hole were also noticed. It has a flat bottom with an axial hole in the centre. Another ring stone was made of basalt and perfectly dressed with chisel marks on the surface as well as in the hole. It tapers towards the upper side. The remaining objects are made

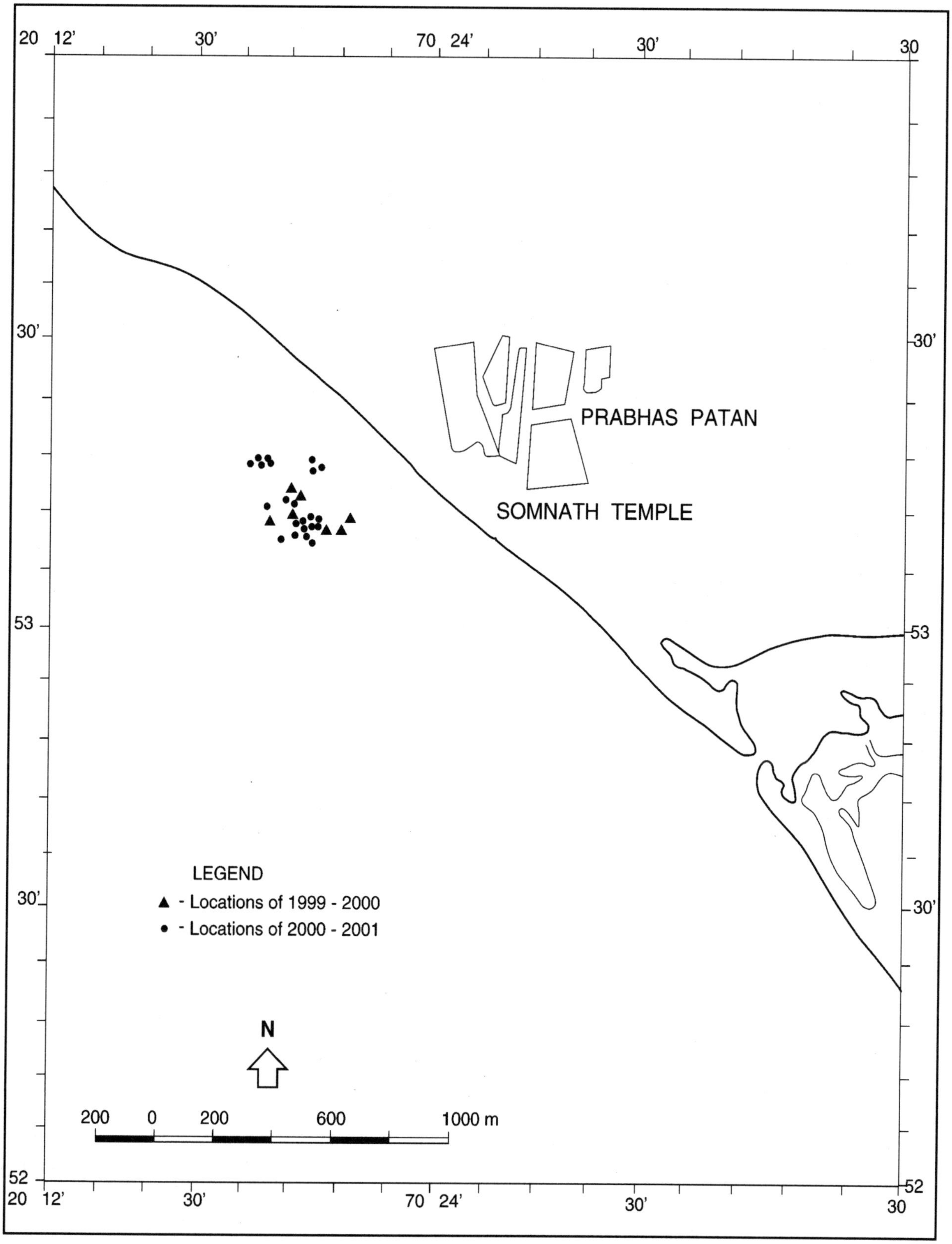

Fig. 21. Location of archaeological artefacts off Somnath.

out of limestone and some of them have been eroded significantly and their axial hole has been widened. A few of them got integrated with bedrock. They are lying scattered and do not suggest any pattern for placing them on the seabed.

Four triangular anchors were also found in this zone and only two are intact while the other two have lost their original shape. One triangular anchor lying exposed on the seabed is the biggest one found from Somnath. It has a perfect triangle shape and even thickness. As usual it has an upper circular hole and two square holes on the lower side. This anchor has been retrieved. Another anchor is very small in size and has three round holes in a triangle. This anchor also also an additional hole in the centre as of Bet Dwarka anchor. This is also made of limestone. The other two anchors are broken and they are made of limestone.

Only one grapnel anchor was noticed and this is also made out of limestone. It has a very rough surface and blunt edges. Only one lower broken hole is present and the upper hole is absent.

The most important discovery from Somnath waters is a marble statue of a goddess. This is lying at 13 m water depth. Perhaps this was made out of marble stone; however, the pedestal is made of some other stone (may be sandstone). The statue appears to be of Goddess Lakshmi and perhaps was thrown in the sea when it was broken. It may be of very recent origin. The statue was retrieved during 2000-01 surveys.

2.3 ZONE 3

The water depth ranges from 6 m to 9 m. The seabed consists of a sandy bottom with a big patch of rock over which beautiful soft corals were noticed. Total eight ring stone objects were recorded, all of them made out of limestone. Among them one ring stone was broken and only half of its portion survives. Four ring stones have a common shape with a flat circular bottom and tapers upper side with an axial hole in the centre. The other three ring stones have a flat, rough and porous surface on both ends. The axial hole has been widened due to erosion. All of them are lying exposed in-between and over the rocks.

Apart from the above archaeological artefacts, a few more ring stones have been recorded at different places in Somnath waters. Two of them are lying in-between rocks in a water depth of 9 m. They are also made out of limestone and have an excel hole in the centre. Another object is lying in 14 m water depth and is trapped in-between two rocks. The object is made out of haematite rock. It is lying tilted and has an excel hole in the centre. One broken triangular anchor was also noticed in a channel 15 m water depth. The upper portion has been broken and the lower two square holes are intact. This is also made out of limestone.

2.4 DISCUSSIONS

In ancient literature, Somnath has been considered as a contemporary site of Dwarka and according to mythology, Lord Krishna visited this place. Extensive explorations off Somnath brought to light a large number of archaeological artefacts, which include 35 ring stone and 6 triangular type and one grapnel type anchor (Figs. 22-23, Pl. 52-53). However, no other structures were reported from Somnath waters. These anchors are similar to those found from Dwarka waters. A detailed discussion has already been given on triangular (composite) and grapnel (Indo-Arab) type anchors in the book on Dwarka; hence the discussion here is limited to the description of artefacts from Somnath waters.

The explored area in Somnath water is located at a distance of 400 m south-west of the Somnath temple. The study area is comprised of rocky outcrop and sandy patches. The first season yielded 6 objects, the second season 8, and the present season (2001) yielded 21 ring

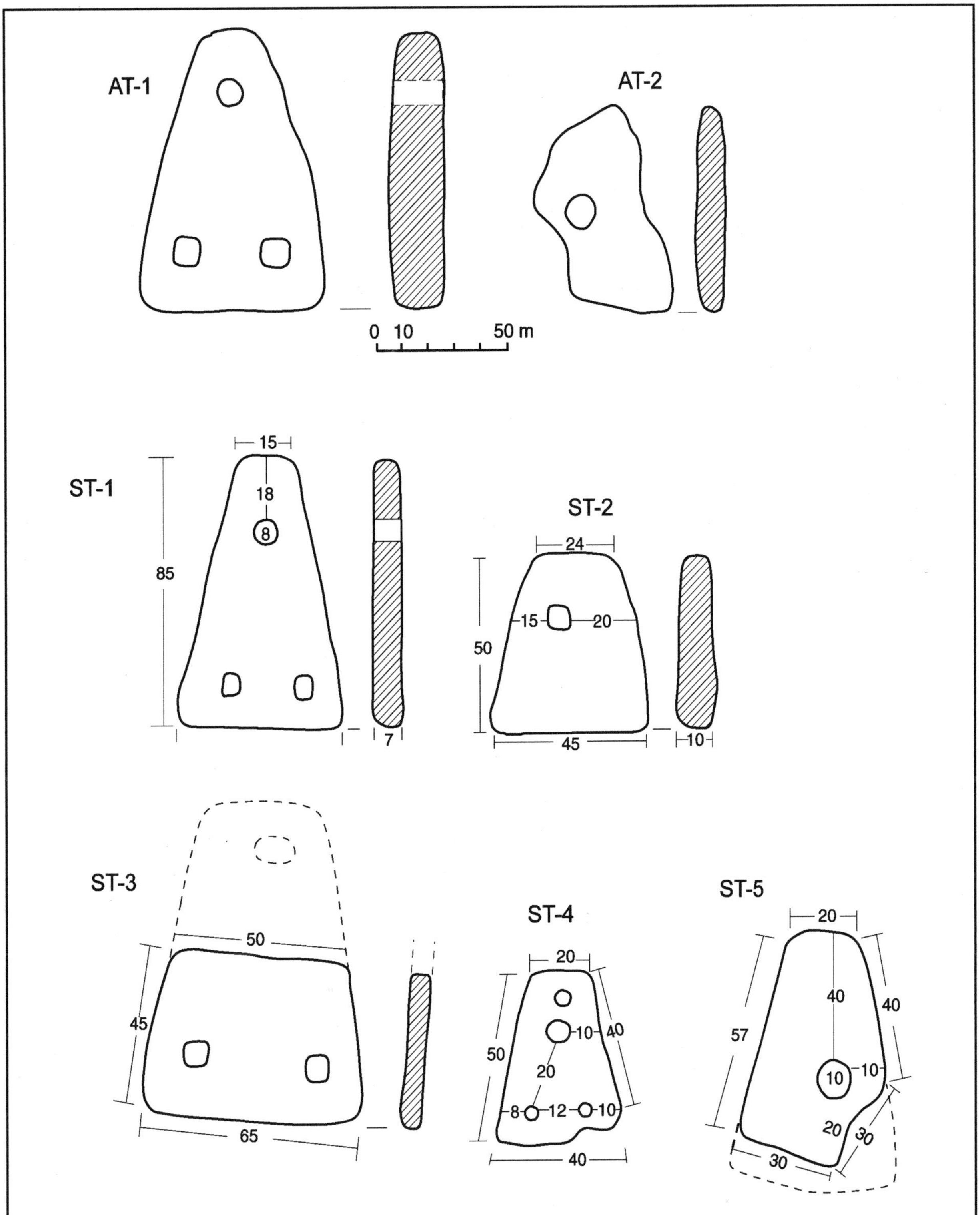

Fig. 22. Composite type anchors from Aramda and Somnath.

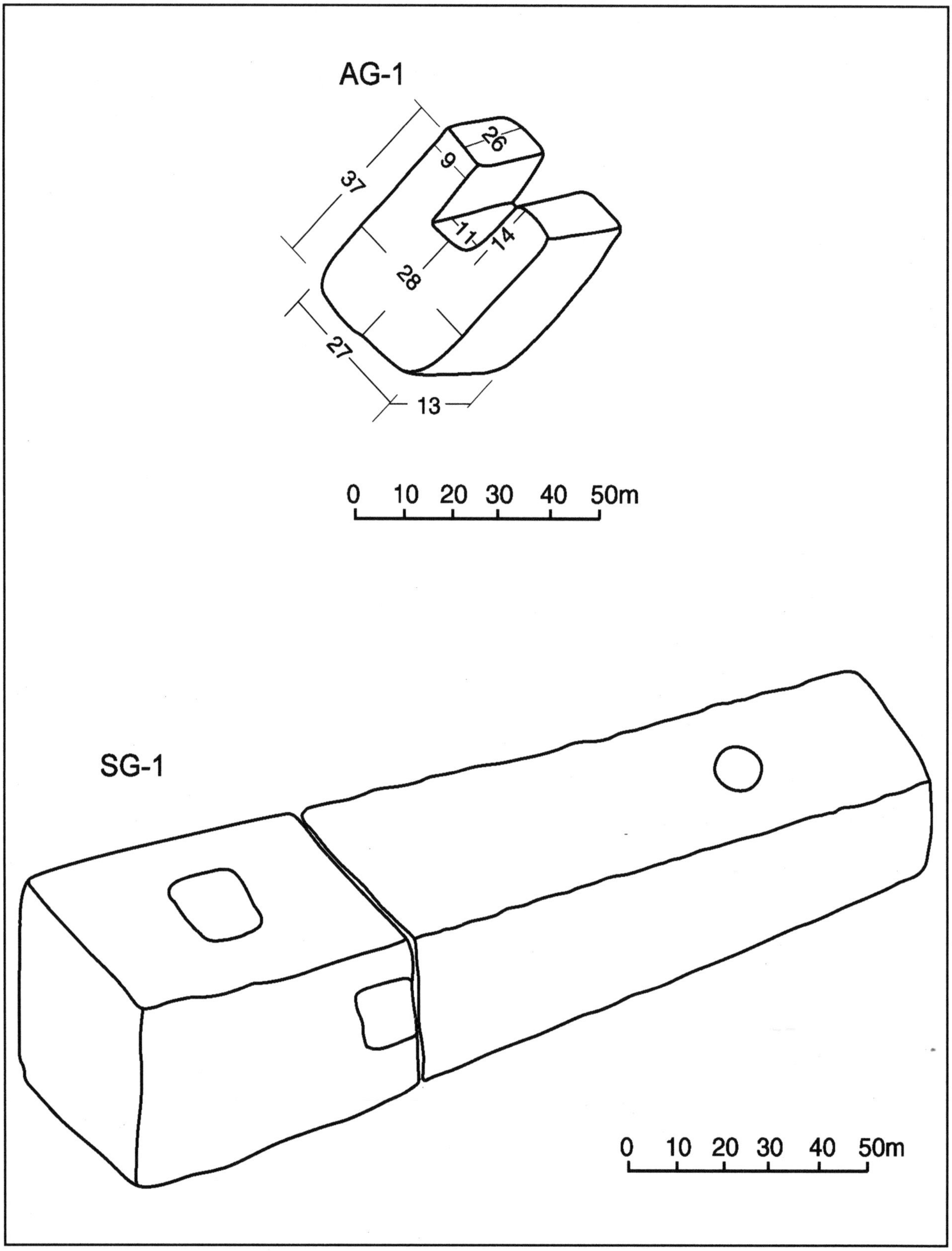

Fig. 23. Indo-Arab type anchors from Aramda and Somnath.

Pl. 43. An old dockyard at Aramda.

Pl. 44. Shiva temple of Somnath overlooking the Arabian Sea.

Pl. 45. A narrow channel in Somnath water observed at 8-10 m water depth.

Pl. 46. A wide channel filled with fine sand in Somnath waters.

Pl. 47. Ring stone anchor lying off Somnath.

SR-1

SR-2

SR-3

SR-4

SR-5

SR-7

SR-8

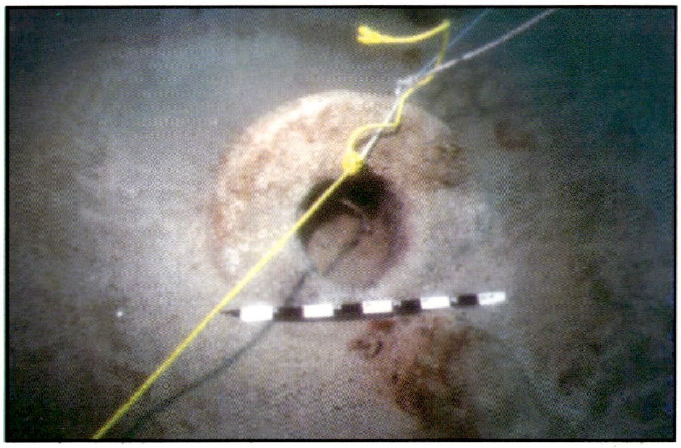
SR-9

Pl. 48. Ring stone anchors No. 1 to 9 from Somnath waters.

SR-10

SR-11

SR-12

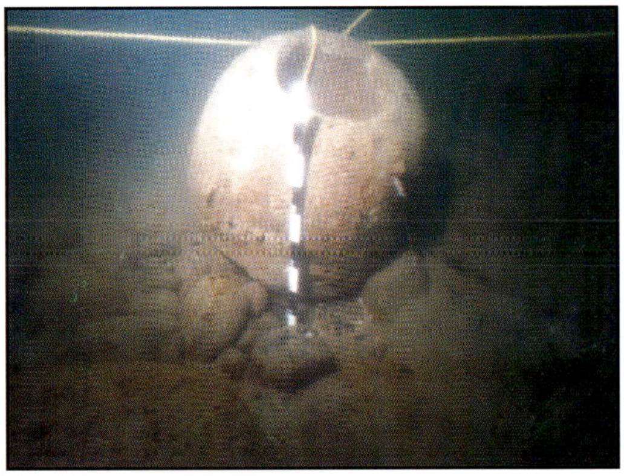

SR-13

SR-14

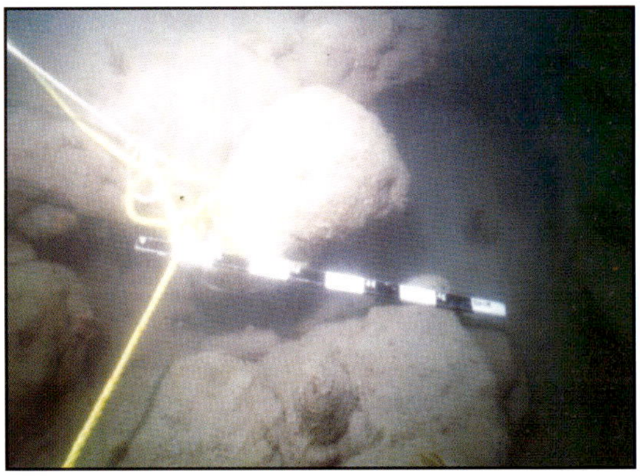

SR-15

SR-16 & 17

SR-18

Pl. 49. Ring stone anchors No. 10 to 18 from Somnath waters.

SR-19

SR-20

SR-21

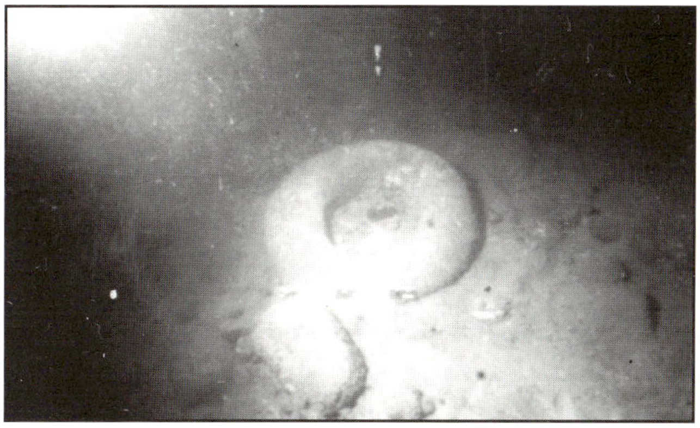

SR-22

SR-23

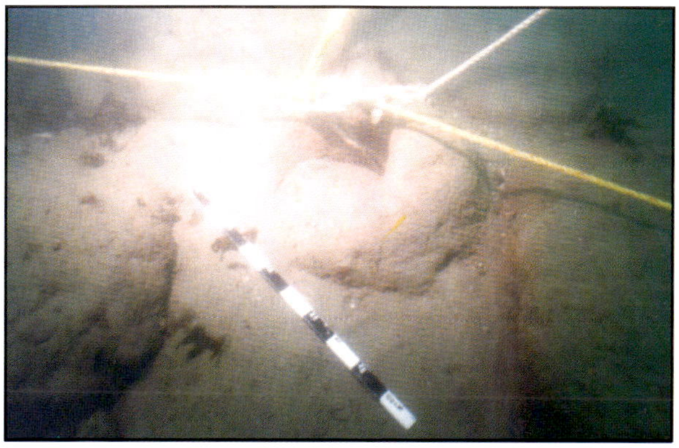

SR-24

SR-25

SR-26

Pl. 50. Ring stone anchors No. 19 to 26 from Somnath waters.

SR-27

SR-28

SR-29

SR-30

SR-31

SR-32

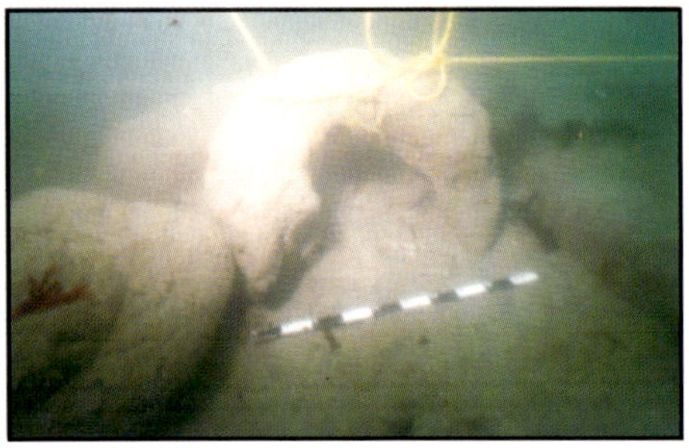
SR-33

SR-34

SR-35

Pl. 51. Ring stone anchors No. 27 to 35 from Somnath waters.

ST-1

ST-2

ST-3

ST-4

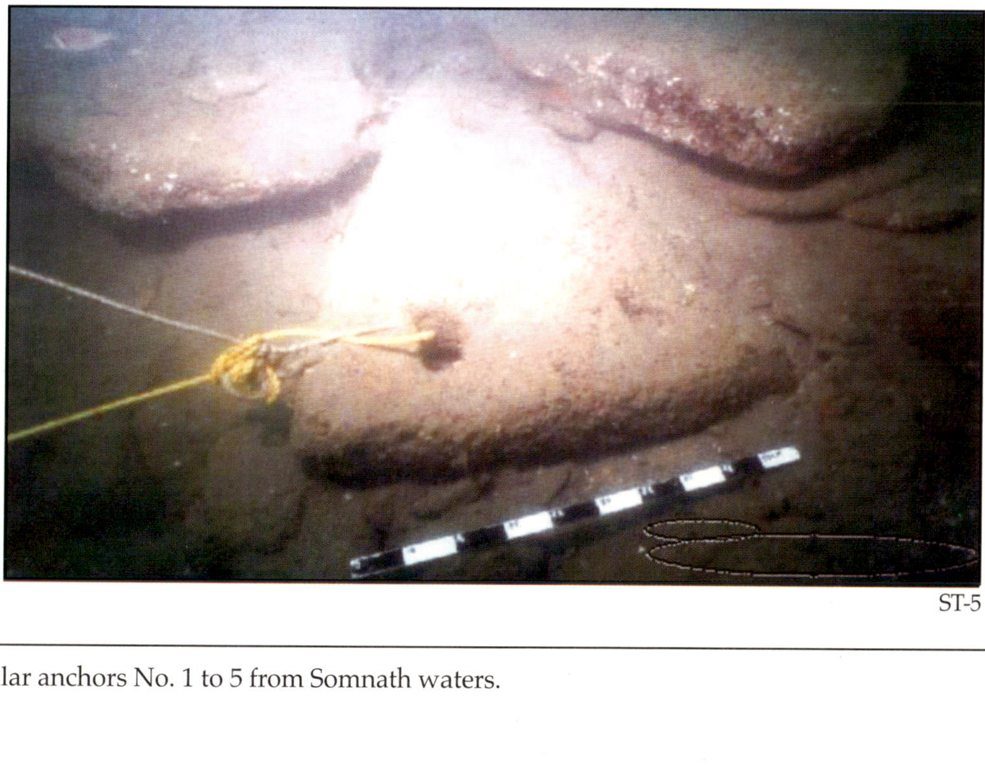

ST-5

Pl. 52. Triangular anchors No. 1 to 5 from Somnath waters.

SG-1

SG-2

Pl. 53. Indo-Arab anchors 1 & 2 from Somnath waters.

stones. Somnath waters witnessed the occurrence of the largest number of ring stones. Out of the 43 stone anchors discovered, 80% belong to the ring stone category. The water depth varies from 7 m to 15 m. The ring stones were noticed in less than 8 m water depth and growth of seaweed was also seen; however, ring stones found in deeper depths have a layer of greyish marine growth. The detailed measurements and an estimated weight of each ring stone is given in Table no. 2. On the basis of shapes and raw materials, the ring stones have been divided into three groups.

2.4.1 Group I

Ring stone nos. 1, 2, 8, 11-13 and 18 are made out of basalt. The majority of them are lying exposed on the seabed; however, a few of them are lying partially buried under sediment. The exposed portions of these ring stones are covered with greyish marine growth. They are spheroid in shape with flat bottom. The upper edge of the axial hole is sharp while the lower edge is blunt. Chisel marks are also noticed on the exterior as well as in the hole. Often they are very heavy and the average estimated weight of ring stones of this group is 220 kg. The heaviest ring stone weighs 392 kg and the lightest 114 kg.

2.4.2 Group II

Ring stone nos. 3, 6, 9-10, 14-16, 19, 21-26, 28-32 and 34 are made out of two types of rock, i.e., limestone and sandstone. They are spheroid in shape with a wide axial hole. Some of the ring stones of limestone are very rough and have been integrated with the seabed. The majority of ring stones are lying exposed on the seabed and a few are lying partially buried in sand. The edge of the upper end of axial hole is sharp and the lower end is blunt. A few ring stones of sandstone have chisel marks in axial hole. The average estimated weight of the ring stones of this group is 82 kg. The heaviest ring stone weighs 221 kg and the weight of the smallest ring stone is 36 kg.

2.4.3 Group III

The remaining ring stone nos. 4, 5, 7, 17, 20, 27, 33 and 35 are made out of very porous limestone. The ring stones of this category are irregular, circular in shape with a low height. The axial hole is irregular and wide. Each ring stone has a rough surface and blunt edges. Most of them are lying exposed on the seabed, out of which one is broken. The estimated average weight of these ring stones is 54 kg. The weight of the heaviest ring stone is 122 kg and the smallest ring stone weighs 16 kg.

2.4.4 Description of ring stones (Figs. 24-27; Pls. 48-51)

1. The anchor was observed partially buried in the sediment close to the edge of a rocky patch in 9 m water depth. This is one of the biggest anchors found in Somnath waters. The anchor was exposed with the help of airlift operation. It has a flat base and gentle tapering towards the apex. An axial hole with chisel marks is also noticed. The anchor has a smooth surface and blunt edges and is made out of basalt.
2. The ring stone anchor was found near a rocky outcrop at a water depth of 8 m. It is made out of basalt. The anchor has a smooth surface and sharp edges. It tapers towards the apex with an axial hole. It has a flat bottom. A thin layer of marine growth is also noticed here.
3. This is comparatively smaller than the earlier two anchors and was lying exposed on the rocky seabed. It is made of limestone. It has a smooth surface and sharp edges. Chisel marks are also noticed. The anchor is covered with seaweed growth.

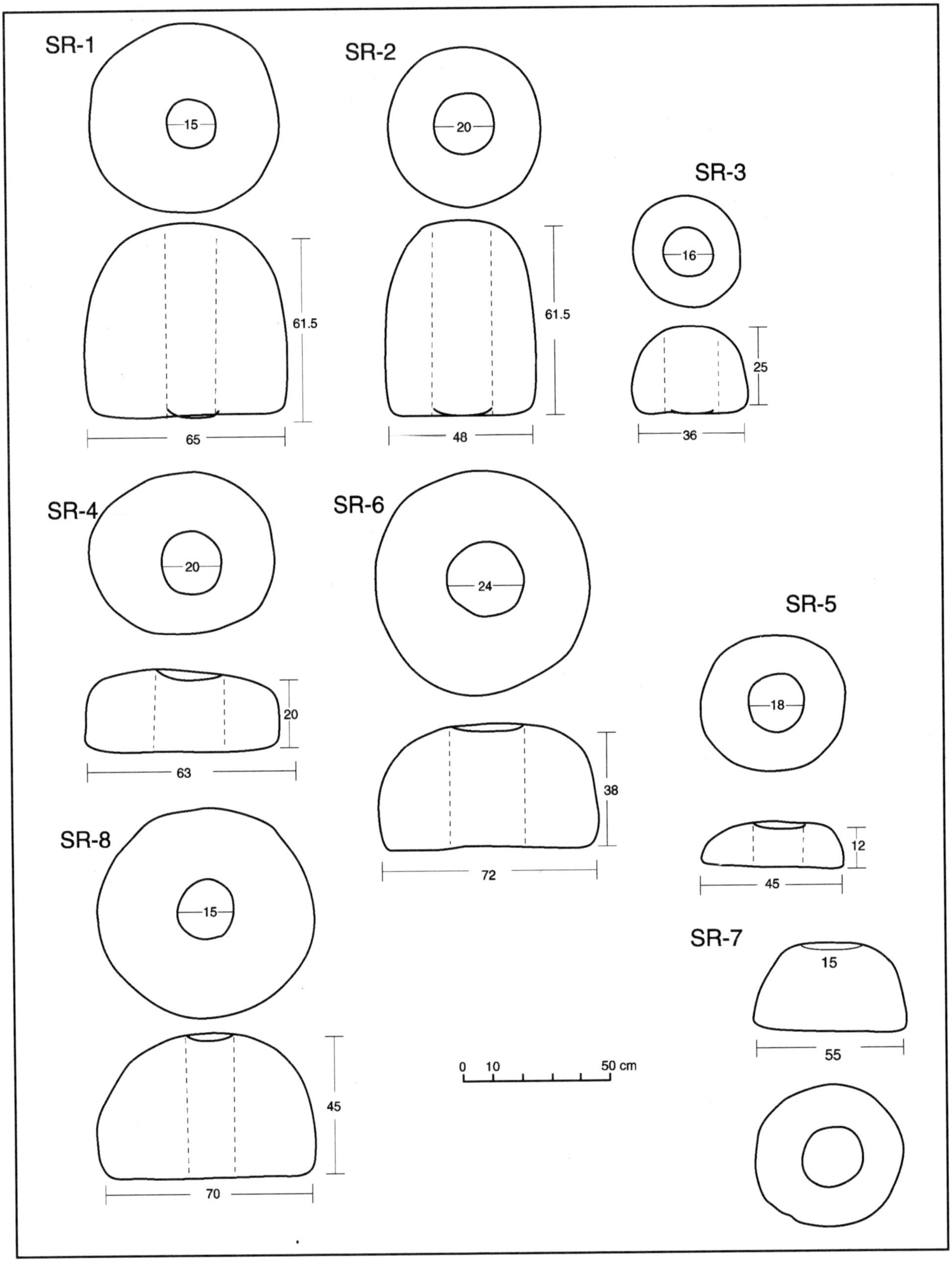

Fig. 24. Ring stone anchors No. 1 to 8 from Somnath waters.

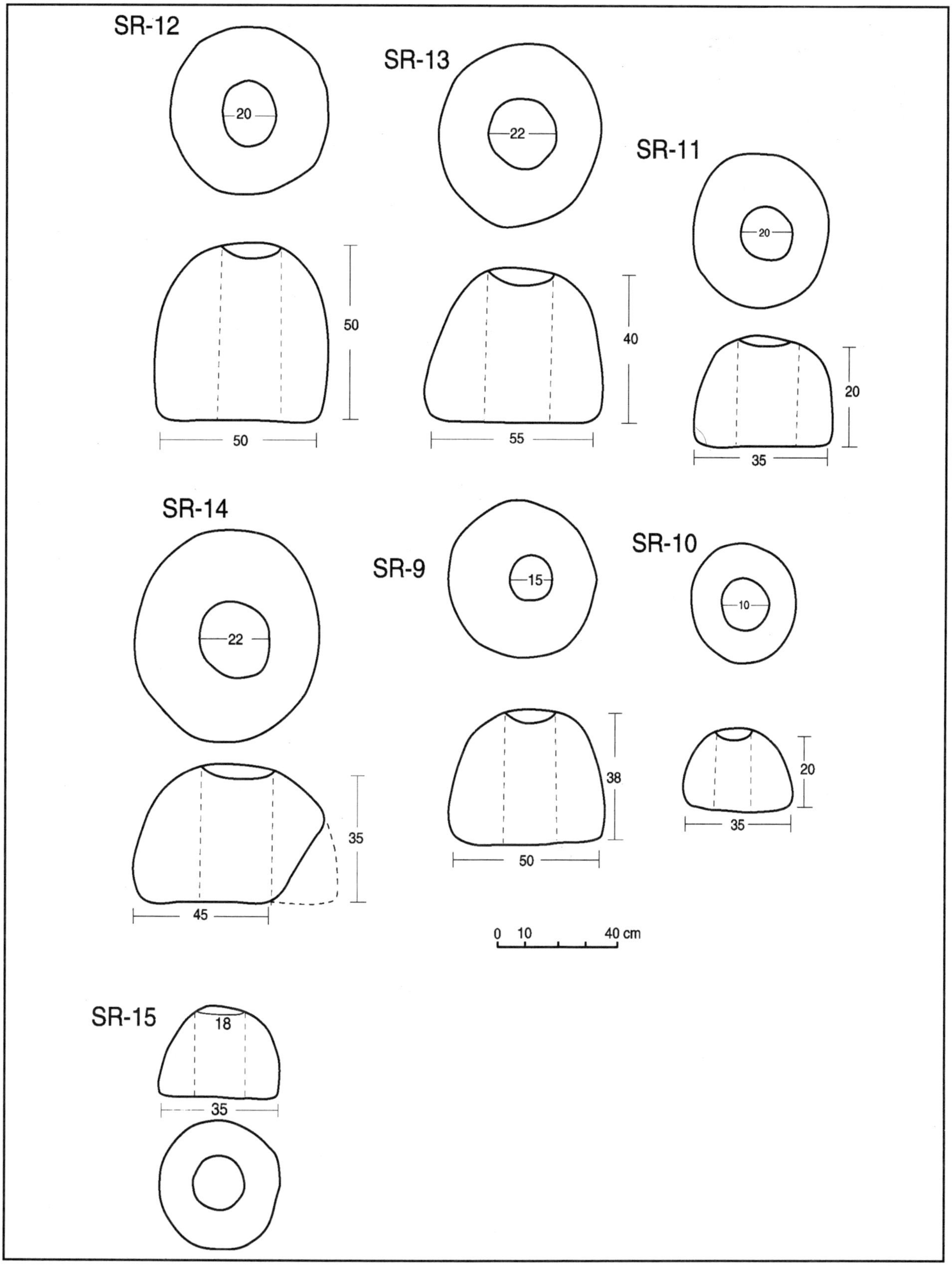

Fig. 25. Ring stone anchors No. 9 to 15 from Somnath waters.

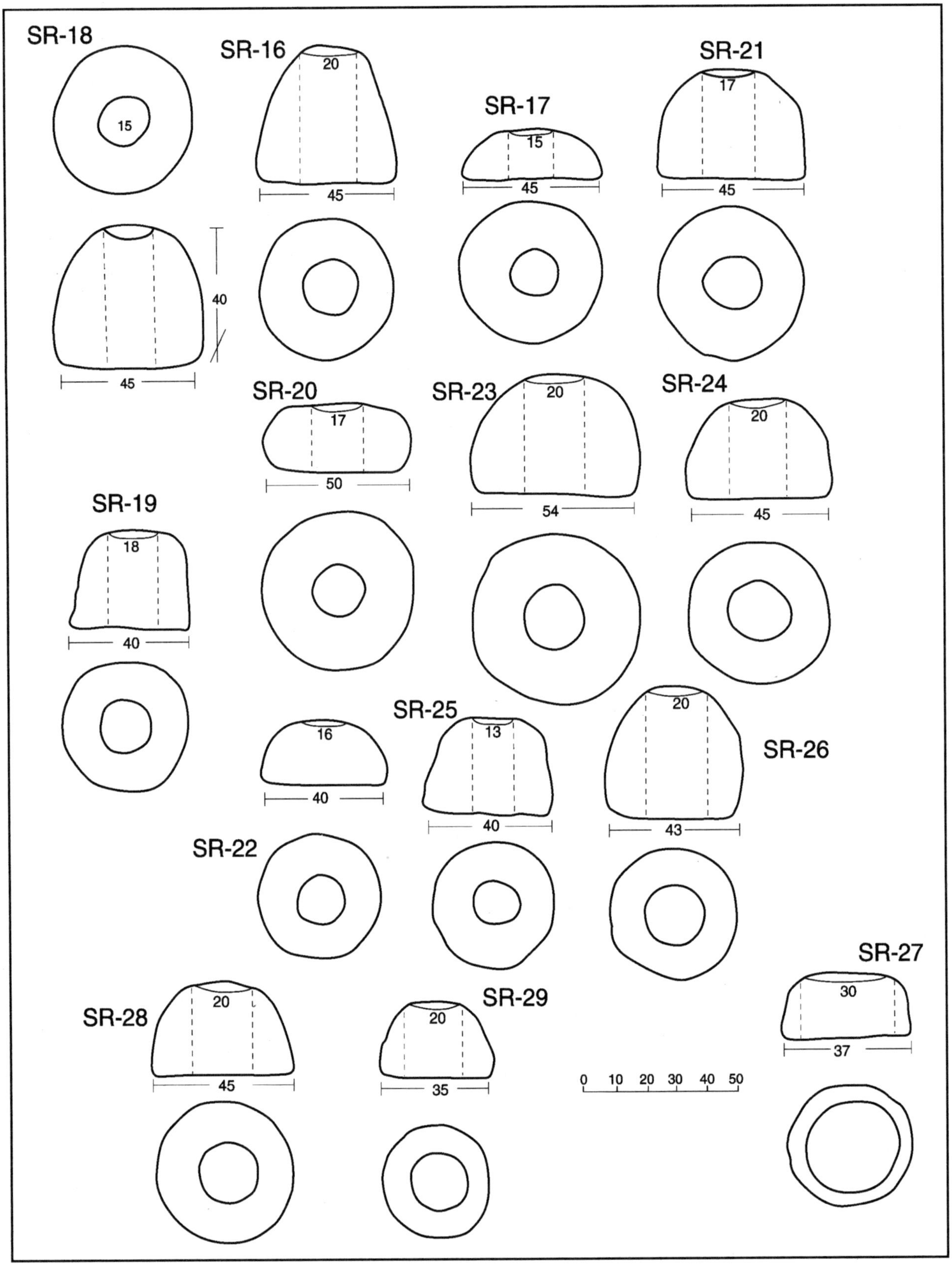

Fig. 26. Ring stone anchor No. 16 to 29 from Somnath waters.

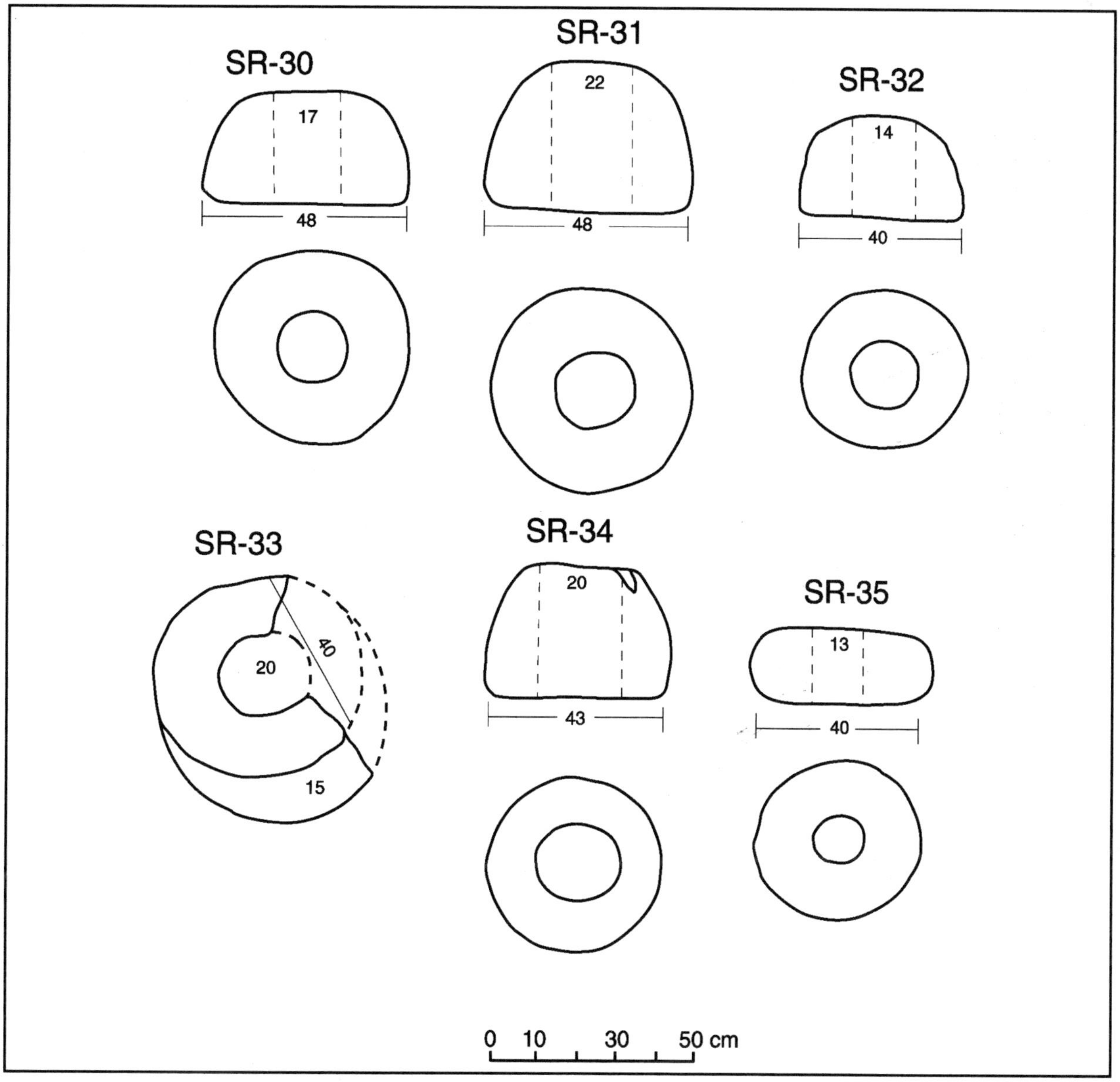

Fig. 27. Ring stone anchor No. 30 to 35 from Somnath waters.

4. The anchor was lying over a rocky patch. It is different from earlier anchors and does not taper towards the apex. This is cut from calcareous rock. It has a very rough surface and irregular shape. The height of this anchor is low.
5. The anchor was observed in a channel, partially buried in sediment. It is made out of a sedimentary rock and has a smooth surface.
6. The ring stone anchor was observed partially buried in sediment and is made out of sandstone. The axial hole has been widened perhaps due to its prolonged use. It has a smooth surface and blunt edges.
7. This was lying near anchor No. 8. It is made out of limestone having a very rough surface and irregular shape, with an axial hole.
8. The ring stone anchor was lying exposed on the seabed. It is made out of basalt and has sharp edges. The anchor has a flat bottom and tapers towards the apex portion.

9. The ring stone anchor was lying in a channel. It is made out of sandstone. It has a smooth surface and sharp edges. The axial hole has been widened perhaps due to its continuous use. It has a flat bottom and a smooth surface.
10. A small sized ring stone was found exposed over the rocky seabed. It is made out of sandstone and has a smooth surface and sharp edges. The anchor has a flat bottom and an axial hole.
11. The anchor was lying upside down, exposed over the rocky seabed. It has been perfectly chiselled. The object has a very rough surface and sharp edges. As usual it has a flat bottom and an axial hole in the middle.
12. The anchor was lying tilted and exposed with a thick layer of marine growth. It is made out of basalt. The object has a flat bottom and tapering upper side. It has been perfectly shaped forming a spherical form. It has a smooth surface and blunt edges. Chisel marks are visible in the hole and at the bottom of the anchor.
13. This is very similar to No. 7 and is partially buried in sand. It is also made out of basalt and has a very smooth surface and sharp edges. As usual it tapers toward the upper side. The object is lying vertically on the seabed.
14. The anchor was partially buried in sediment and is made out of limestone. The exterior of the anchor has been eroded. It is lying exposed in a channel. As usual it has a flat bottom and tapers on the apex side.
15. The ring stone was found partly buried in sediment and also trapped in-between rocks. A thick layer of marine growth is also observed. The anchor is made out of limestone and has a rough surface. The upper end of the hole has been widened. The exterior of the object is uneven.
16. The ring stone anchor was lying exposed and tilted. This is made out of limestone. A thick layer of marine growth was noticed. It has an even and rough surface. Chisel marks have also been noticed.
17. The ring stone was found exposed on the seabed. It is made out of limestone. It has an irregular shape and low height. It is not tapering on the upper side as other anchors. It has a rough surface and blunt edges. A layer of marine growth is also noticed. The upper and lower ends of the hole have been widened.
18. The ring stone anchor was lying exposed over the sandy seabed. It is made out of basalt. This is one of the most perfect anchors found from Somnath. The edges are sharp and the surface is smooth. A grey layer of marine growth is also observed. Chisel marks are also noticed in the hole.
19. The ring stone was lying partly buried in sediment. The exposed portion is covered with marine growth. It has been eroded and the upper surface is uneven. The lower portion of the anchor is integrated with bedrock. The upper side of the hole has been widened.
20. Limestone ring stone was observed trapped in-between rocks. It has a very irregular shape and rough surface. A thick layer of marine growth is also noticed. The bottom of the object is uneven.
21. Ring stone anchor was lying partly buried in the sediment. It is made out of limestone and has an irregular shape. The hole on the upper side has been widened. The exposed portion is covered with marine growth.
22. The ring stone anchor was trapped in-between the rocks, made out of limestone. It has a rough and porous surface. The upper end of the axial hole has been widened. A layer of marine growth is noticed.

23. The ring stone anchor was lying exposed on the rocky seabed. It has a rough and uneven surface. Chisel marks in the hole have also been observed. A layer of marine growth is noticed. It is made out of limestone.
24. The anchor was found partially buried in sediment. It is made out of limestone and has a very fine shape and smooth surface. The lower side of the hole has been widened while the edge of the upper side is sharp. Chisel marks are also noticed in the hole.
25. The anchor was found exposed on the rocky seabed and is made out of limestone. It has a rough surface and blunt edges. The upper and lower side of the axial hole have been widened. A thick layer of marine growth has been noticed.
26. The ring stone anchor was lying exposed on the rocky seabed and has been cut out of laterite. It has fine finish and blunt edges. This has a rough and porous surface. The upper edge of the hole is sharp while the lower edges are blunt and the hole has been widened. A thick layer of marine growth is also noticed.
27. The limestone anchor was noticed on the rocky seabed. The anchor has a wide hole and low height. It has a very smooth surface with sharp edges. A layer of marine growth is also observed.
28. The ring stone anchor was observed partly buried in sediment and is made out of hard red stone (red sandstone or haematite). It has a smooth surface and blunt edges. Chisel marks in the hole are also noticed. The exposed portion is covered with marine growth.
29. The ring stone was found partly buried in sediment and is made out of fine grained limestone; however, the exterior of the anchor has been eroded. It has an uneven and rough surface. The exposed portion is covered with marine growth. The axial hole has been widened due to continuous use.
30. The limestone anchor was lying exposed on the seabed. The surface of the object is also eroded. Lower side hole has been widened, perhaps due to continuous use for a long time. It has blunt edges and a layer of marine growth is also noticed.
31. The ring stone anchor was noticed partially buried in the sediment. It is made out of sandstone. It has a smooth surface and blunt edges. The lower side of the hole has been widened perhaps due to continuous use. Chisel marks are also noticed in the hole. Marine growth is also observed.
32. The ring stone anchor was lying exposed on the rocky seabed. It has an irregular shape with rough and uneven surface. This is made out of limestone. The upper side of the hole has been widened. Marine growth has been noticed over the exposed portion of the anchor.
33. The ring stone anchor was lying exposed on the sandy seabed and is made out of limestone. The anchor has a low height and a wide axial hole. It has a rough and porous surface with blunt edges. Marine growth is also observed.
34. The ring stone anchor was lying tilted between the boulders and it is made out of limestone. It has an irregular shape. The anchor has a rough, porous and uneven surface. The upper end of the hole has been widened. It has blunt edges and a layer of marine growth is noticed.
35. The anchor was lying exposed on the rocky seabed and is made out of limestone. It has an irregular shape. The anchor has a very rough and porous surface. No proper edging and cutting is seen. The ends of the hole have been widened and also a cutting over the edge is significantly visible. Marine growth is also observed. The surface of the anchor is uneven.

TABLE NO. 4. RING STONES FROM SOMNATH WATERS
(DIMENSIONS IN CENTIMETRE)

Sl. no.	Year of find	Raw material	Max. diameter (cm)	Diameter of hole (cm)	Height (cm)	Weight in kg (estimated)
1	1992	Basalt	65	15	61.5	392
2	1992	Basalt	48	20	61.5	198
3	1992	Limestone	36	16	25	37
4	1992	Limestone	63	20	20	122
5	1992	Sandstone	45	18	12	25
6	2000	Sandstone	72	24	38	221
7	2000	Limestone	35	15	20	70
8	2000	Basalt	70	15	46	293
9	2000	Sandstone	50	15	38	109
10	2000	Limestone	35	11	20	31
11	2000	Basalt	35	20	20	114
12	2000	Basalt	50	20	50	167
13	2000	Basalt	55	22	40	162
14	2000	Limestone	45	20	35	81
15	2001	Limestone	35	18	20	36
16	2001	-do-	45	20	40	113
17	2001	-do-	45	15	13	40
18	2001	Basalt	45	15	40	118
19	2001	Limestone	40	18	30	66
20	2001	-do-	50	17	20	77
21	2001	Limestone	45	17	25	43
22	2001	Limestone	40	15	20	48
23	2001	-do-	54	20	37	162
24	2001	Limestone	45	20	30	85
25	2001	Limestone	40	13	34	85
26	2001	Laterite	43	20	36	89
27	2001	Limestone	37	30	20	16
28	2001	Haematite	45	20	28	75
29	2001	Limestone	35	20	25	36
30	2001	Limestone	48	17	26	91
31	2001	Sandstone	48	22	38	111
32	2001	Limestone	40	14	18	44
33	2001	Limestone	48	20	15	46
34	2001	Limestone	43	20	30	76
35	2001	Limestone	40	13	15	37

3
Conclusion

Like Dwarka, Somnath has been an important place for Hindu pilgrims as a grand temple dedicated to Lord Shiva overlooks the Arabian Sea. Somnath also finds place in the ancient Indian literature. Land antiquity of Somnath/Prabhasa dates back to the late 3rd and early 2nd millennium BC. Also, remains of the Historical Period have been recorded.

Underwater explorations have yielded a large number of stone anchors, which include composite, grapnel and ring stone type. However, ring stone anchors are in the majority among the anchors as they comprised of more than 80% of the total anchors recovered from Somnath. The findings of such a large number inclined to suggest that ring stone anchors might have originated in this region and, therefore, Gaur *et al.* (2002) have described them as Saurashtra stone anchors. The raw material used for these anchors (basalt and limestone) also supports this hypothesis. The findings of a large amount of anchors also indicated that Somnath was an active port and a safe anchorage during the Historical and the Medieval periods.

Marine archaeological explorations on the Saurashtra coast indicate that ancient anchoring points fall between 5 to 13 m water depths. At Somnath the anchoring point was between 9-13 m water depth. Perhaps anchoring points are well connected with the tidal range at respective places. For instance, tidal fluctuation is very high at Dwarka and Bet Dwarka (between 3 to 4.5 m) and ancient anchoring point falls at 5 to 7 m whereas at Veraval tidal fluctuation is low (less than 2.5 m) which might have necessitated anchoring point at a deeper depth.

Offshore findings are similar to those found from Dwarka which have been dated between the Historical and the Medieval periods. Therefore, a similar date for underwater findings from Somnath may also be assigned.

References and Bibliography

Anonymous, 1884. *Gazetteer of Bombay Presidency*, Kathiawad.

Ansari, Z.D. and M.S. Mate, 1966. *Excavations at Dwarka,* Post Graduate Research Institute, Deccan College, Pune.

Athiyaman, N., 2001. 'Stone Anchors from East Coast of India: A Precursor for Marine Archaeology of Tamilnadu, in *Historical Archaeology, Epigraphy, Numismatics and Archives*, Proceedings of 61st Session of Indian History Congress.

Bass, G.F., 1967. 'Cape Gelidonya: A Bronze Age Shipwreck', *Transaction of American Phil. Society,* 57.8.

——, 1976. 'Sheytan Desesi: Preliminary Report', *IJNA,* 5.4:293-303.

——, 1986. 'Bronze Age Shipwreck at Ul Burun (Kas): 1984 Campaign', *AJA,* 90.3:269-96.

Bhatt, N. and M.P. Patel, 1998. 'Bioclastic Shore Deposits: Indicators of Late Quaternary High Sea in Saurashtra, Western India', *JGSI,* 52:537-42.

Bhatt, N., 2000. 'Lithostratigraphy of the Neogene-Quaternary Deposits of Dwarka-Okha Area, Gujarat', *JGSI,* 55:139-148.

Bisht, R.S., 1991. 'Dholavira: A New Horizon of the Indus Civilization', *Puratattva,* 20:71-82.

Chandra, M., 1977. *Trade and Trade Routes in Ancient India,* Abhinav Publications, New Delhi.

Chittick, N., 1980. 'Stone Anchor Shanks in Western Indian Ocean', *IJNA,* 9.1:73-6.

Dhavalikar, M.K., M.R. Raval and Y.M. Chitalwala, 1996. *Kuntasi – A Harappan Emporium on West Coast,* Pune.

Dimitrov, B., 1976. 'Stone Anchors from Sozopol Bay', *IJNA,* 5:81-83.

——, 1977. 'Anchors from the Ancient Ports of Sozopol', *IJNA,* 6:156-163.

Eiseman, C.J., 1973. 'Amphoras from the Porticellow Shipwreck (Calabria)', *IJNA,* 2.1:13-24.

Epigraphia Indica, 1981. Archaeological Survey of India, New Delhi, Vol. VIII: 59-96.

Epigraphia Indica, 1981. Archaeological Survey of India, New Delhi, Vol. VII: 47-74.

Fairbank, R.G., 1989. 'A 17,000 Years Glacio-Eustatic Sea Level Record: Influence of Glacial Melting Rates on the Younger Dryas Event and Deep Ocean Circulations', *Nature,* 342:637-642.

Fairbridge, R.W., 1960. 'The Changing Level of the Sea', *Scientific American,* 202.5:70-79.

Fedden, F., 1884. 'Geology of Kathiawar Peninsula in Guzerat', *Mem. Geol. Survey of India,* V. 21-22:78-125.

Frey, D., F.D. Hentschel and D.M. Keith, 1978. 'Deep Water Archaeology: The Capistello Wreck Excavation, Lipari, Aeolion Island', *IJNA,* 7.4:279-300.

Frost, H., 1970. 'Bronze Age Stone Anchors from the Eastern Mediterranean', *MM,* 56.4:377-94.

——, 1979. 'Egypt and Stone Anchors: Some Recent Discoveries', *MM,* 65.2:137-161.

——, 1991. 'Anchors', *Ras Shamra-Qugarit*, VI, ERC, Paris.

Frost, H., W. Culicon and J.E. Curtis, 1974. 'The Punic Wreck in Sicily: Second Season of Excavation', *IJNA*, 3. 1:35-54.

Gaur, A.S., 1990. 'Ceramic Industries of Dwarka', *JMA*, 1:110-113.

——, 2000. *Harappan Maritime Legacies of Gujarat*, Asian Publication Services, New Delhi.

Gaur, A.S., Sundaresh, P. Gudigar and S. Tripati, 1994. 'Protohistoric Ceramics of Dwarka and Bet Dwarka', *The Role of Universities and Research Institutes in Marine Archaeology*, S.R. Rao (ed.), NIO, Goa, pp. 165-171.

——, 1997. 'An 18th Century Shipwreck off Poompuhar Tamil Nadu Coast (India)', *IJNA*, 26.2:118-126.

Gaur, A.S. and Sundaresh, 1998. 'Underwater Exploration off Poompuhar and Possible Causes of its Submergence', *Puratattva*, 28:84-90.

——, 2007. 'Evidence of Shoreline Changes in the Gulf of Kachchh, Based on the Submerged Temple Complex at Pindara', *CS*, 92.6:733-35.

Gaur, A.S. and K.H. Vora, 1999. 'Ancient Shorelines of Gujarat during Indus Civilization (Late Mid-Holocene Period): A Study Based on Archaeological Evidences', *CS*, 77.1:180-5.

Gaur, A.S., Sundaresh, Sila Tripati, P. Gudigar, K.H. Vora and S.N. Bandodkar, 2001. 'A Group of 20 Stone Anchors from the Waters of Dwarka, on the Gujarat Coast, India', *IJNA*, 30.1:95-108.

Gaur, A.S., Sundaresh, Sila Tripati, and S.N. Bandodkar, 2002 . 'Ring Stones from Gujarat Coast, West Coast of India', *MM*, 88.4:390-404.

Gaur, A.S., Sundaresh and Sila Tripati, 2004. 'Stone Anchors (Composite Type) from the Saurashtra Coast, India: An Indicator of Ancient Ports and Sea Routes', *BAIMA*, 28:125-132.

——, 2004. 'Grapnel Anchors from Saurashtra: Remnants of Indo-Arab Trade Route in Indian Ocean', *MM*, 90.2:134-151.

——, 2005. 'A Late Harappan Port at Kindar Kheda', *ME*, 30(2):44-48.

——, 2005. 'Maritime History of Saurashtra-Kathiyawad Coast: Study Based on the Marine Archaeological Investigations', *RASB*, 50:159-195.

——, 2005. 'Ancient Dwarka: Study Based on Recent Underwater Archaeological Investigations', *Migration and Diffusion*, 6.21:56-77.

——, 2006. 'Evidence for Indo-Roman Trade from Bet Dwarka Waters, West Coast of India', *IJNA*, 35.1:117-127.

——, 2006. 'Marine Archaeological Explorations on the Southwestern Coast of Saurashtra, India', *JIOA*, 3: 81-89.

Gaur, A.S., Sundaresh and K.H. Vora, 2005. *Archaeology of Bet Dwarka Island: An Excavation Report*, Aryan Books International, New Delhi.

——, 2007. 'Shoreline Changes During Last 2000 Years on the Saurashtra Coast of India: Study Based on Archaeological Evidences', *CS*, 92.1:103-110.

Ghose, A., 1989. *An Encyclopaedia of Indian Archaeology*, Delhi, pp. 348-350.

Gokani, P., 1988. 'Marine Archaeology Entering a New Era of Ancient History with Special Reference to Dwarka of Beloved Lord Krishna', *MAIOC*, S.R. Rao (ed.), NIO, Goa, pp. 17-18.

Gopal, L. 1970, 'Indian Shipping in Early Medieval Period', in *India's Contribution to World Thought and Culture*, Vivekanand Commemoration Volume, Lokeshchandra (ed.), Madras, pp. 108-122.

Grossmann, E. and S.A. Kingsley, 1996. 'A Three-Hole Stone Anchor with Wood Remain from Crusadar Arsuf (Apollonia) Israel', *IJNA*, 25.1:49-54.

Gupta, S.K., 1972. 'Chronology of Raised Beaches and Inland Coral Reefs of the Saurashtra Coast', *Journal of Geology*, 80:357-361.

Hegde, K.T.M., K.K. Bhan, V.H. Sonawane, K. Krishnan and D.R. Shah, 1990. *Excavation at Nageswar-Gujarat*, M.S. University, Baroda.

Hora, S.L., 1954. 'Angling in Ancient India', *Ancient India*, No. 10-11, Archaeological Survey of India, New Delhi.

Hourani, G.F., 1975. *Arab Seafaring in the Indian Ocean and Early Medieval Times*, New York.

Indian Archaeology – A Review, 1971-72, Archaeological Survey of India, New Delhi, p. 12.

Indian Archaeology – A Review, 1975-76, Archaeological Survey of India. New Delhi, p. 13.

Indian Archaeology – A Review, 1979-80, Archaeological Survey of India, New Delhi.

Iyengar, R.N. and B.P. Radhakrishnan, 2005. 'Evolution of the Western Coastline of India and the Probable Location of Dwarka of Krsna: Geological Perspective', *JGSI*, 66(3):285-292.

Jain, R.L., 1990. 'Tertiary Rocks of Gujarat', *Rec. Geol. Surv. India*, 123/7:36-38.

——, 1997. 'A Study of Miocene Mollusca from Jamnagar District', Gujarat, Ph.D. Thesis, M.L. Sukhadia University , Udaypur (Unpublished).

Jain, R.L. and R.K. Agarwal, 1992. 'Tertiary Rocks of Gujarat', *Rec. Geol. Surv. India*, 124/7:26-29.

Jaykumar, P. and N. Athiyaman, 1999. 'Stone Anchors of Tamilnadu Coast', Paper presented in *Second International Conference on Marine Archaeology*, Thane, Mumbai on 8-10 January 1999.

Joshi, J.P., 1990. *Excavation at Surkotada 1971-72 and Exploration in Kutch*, MASI No. 87, Archaeological Survey of India, New Delhi.

Kapitan, G., 1979. 'Laoteria from the Sea', *IJNA*, 8.2:97-100.

——, 1994. 'Stone-sank Anchors of the Arab-Indian Trade Period – Were They Mooring Anchors?', *BAIMA*, 18:1-6.

Kenny, L.B., 1993. 'Archaeology and Krsna's Dvaraka', *The Journal of the Asiatic Society of Bombay*, 64-66: 95-100.

Mackay, E.J.H., 1976. *Further Excavations at Mohenjo-daro*, Indological Book Corporation, New Delhi, pp. 595-597.

Mahabharata, 'Mausalparva', 7.41-42 (Hindi Translation) Geeta Press, Gorakhpur.

Majumdar, N.G., 1934. *Explorations in Sind*, Memoir of Archaeological Survey of India, No. 48.

Marshall, John, 1973. *Mohenjo-Daro and the Indus Civilization*, Indological Book House, New Delhi, p. 61.

Mathur, U.B., K.K. Verma and S. Mehra, 1988. 'Tertiary-Qaternary Straitigraphy of Porbandar Area, Southern Saurashtra, Gujarat', *Special Publ., 11(II), Geol. Surv. of India*, pp. 333-345.

Max Muller, F. (ed.), 1894. *The Sacred Books of the East: The Questions of King Milinda* (translated by T.W. Rhys Davids), Part II, Vol. 36, Reprinted in 1988, Motilal Banarsidass, New Delhi, p. 299.

McCrindle, J.W., 1879. *The Commerce and Navigation of the Erythraean Sea*, Reprinted in 1987 by Eastern Book House, Patna.

——, 1885. *Ancient India Described by Ptolemy*, R.C. Jain (ed.), reprinted in 1985, New Delhi, p. 36.

Merh, S.S., 1992. *Quaternary Sea Level Changes along Indian Coast*, Proc. Indian Nayl. Sci. Academy, 58:461-472.

Mookherji, R.K., 1912. *Indian Shipping: A History of the Sea-Borne Trade and Maritime Activities of the Indians from the Earliest Times*, Longmans, Bombay.

Moreland, W.H., 1939. 'The Ships of the Arabian Sea about AD 1500', *JRAS*, 63-74.

Pandey, D.K., U.B. Mathur, R. Singh and T. Bahadur, 2003. 'Late Pleistocene Coral-Algal Sediments of Chaya Formation of Mithapur Area, Jamnagar', Gujarat, *JGSI*, 61:195-201.

Pargiter, F.E., 1904. *The Markandeya Purana* (English translation), The Asiatic Society, Calcutta.

——, 1922. *Ancient Indian Historical Tradition*, Motilal Banarsidass, Delhi.

Pathak, M.C., K.L. Kotnala, N. Prabaharan and K.M. Sivakholundu, 1988. 'Disastrous Effects of Transgressing Sea over Coastal Area – A Case Study of Dwarka and Bet Dwarka', *MAIOC*, S.R. Roa (ed.), National Institute of Oceanography, Goa, pp. 58-62.

Pusalkar, A.D., 1943. 'Historicity of Krishna', in *The Glory that Was Gurjaradesa*, K.M. Munshi (ed.), Bharatiya Vidya Bhavan, Bombay, pp. 111-127.

——, 1945. 'Dwarka', in *B.C. Law Volume 1*, D.R. Bhandarkar, K.A. Nilkantha Shastri, B.M. Barua, B.K. Ghosh and P.K. Gode (eds.), pp. 218-224.

Raban, A., 1981. 'Recent Maritime Archaeological Research in Israel', *IJNA*, 10.4:287-308.

——, 1990, 'Medieval Anchors from the Red Sea', *IJNA*, 19.4:299-306.

——, 2000. 'Three-hole Composite Stone Anchors from a Medieval Context at Caesarea Maritima, Israel', *IJNA*, 29.2:260-272.

Rajamanickam, G.V., 1992, 'Marine Archaeological Exploration Activities from the Centre of Tamil University, Tamilnadu', *Seminar on Marine Archaeology*, N. Kasinathan (ed.), Madras, 34-44.

Rao, S.R., 1979. *Lothal – A Harappan Port Town*, Memoir of Archaeological Survey of India, New Delhi.

——, 1984. 'Submergence of Dwarka is a Fact', *Jnan Jyothi*, S.R. Rao (ed.), Bangalore, pp. 16-26.

——, 1985. 'Marine Archaeological Explorations Off Dwarka, Northwest Coast of India', *Indian Journal of Marine Sciences*, Vol. 16.

——, 1987. *Progress and Prospects of Marine Archaeology in India*, NIO, Goa.

——, 1988. 'Excavation of Submerged Ports – Dwarka A Case Study', *MAIOC*, S.R. Rao (ed.), NIO, Goa, pp. 47-53.

——, 1988. 'Marine Archaeological Research in India', *40 Years of Research – A CSIR Overview*, N.R. Rajagopal (ed.), New Delhi, pp. 231-253.

——, 1990. 'Excavation of the Legendary City of Dvaraka in the Arabian Sea', *JMA*, 1:59-98.

——, 1991. *Dawn and Devolution of Indus Civilization*, Aditya Prakashan, New Delhi.

——, 1992. 'Further Excavations Off Dvaraka and Bet Dvaraka', *Recent Advances in Marine Archaeology*, S.R. Rao (ed.), NIO, Goa, pp. 51-59.

——, 1999. *The Lost City of Dvaraka*, Aditya Prakashan, New Delhi.

Rao, S.R., S. Tripati, A.S. Gaur, P. Gudigar, Sundaresh and K.H. Vora, 1994. 'Underwater Archaeological Exploration of Dwarka and Somnath – Results of 1992 Survey and Problems to Attend', *RURIMA*, S.R. Rao (ed.), NIO, Goa, pp. 113-119.

Rao, S.R., S. Tripati and A.S. Gaur, 1992. 'A Preliminary Exploration of Prabhasa-Somnath', *JMA*, 3:13-16.

Rao, S.R. and A.S. Gaur, 1992. 'Excavation at Bet Dwarka', *JMA*, 3:42-47.

Ratnagar, S., 1981. *Encounters: The Westernally Trade of the Harappan Civilization*, Delhi.

Rhys Davids, T.W. (trans.), 1894. *The Questions of King Milinda*, Edited by F. Max Muller, *The Sacred Books of the East*, Vol. 36, Reprinted in 1988, Motilal Banarsidass, Delhi.

Sakhawala, M.V. and D.M. Shringarpure, 1990. 'Age of Dwarka Beds and Subsequent Strata at Beyt Shankhodhar, Gulf of Katch, Gujarat State', *Mem. Geol. Soc of India*, 16:111-116.

Sankalia, H.D., S.B. Deo, Z.A. Ansari and S. Ehrhardi, 1960. *From History to Prehistory at Nevasa* (1954-60), Poona.

——, 1966. 'Dwarka in Literature and Archaeology', *Excavations at Dwarka*, Ansari and Mate, Poona.

——, 1974. *Prehistory and Protohistory of India and Pakistan*, Bombay.

Sarkar, H., 1953. 'Fish-hooks from the Indus Valley', *Journal of Asiatic Society, Science*, XIX:133-339.

Shastri, D.S., 1941. *Aitihasik Sansodhan*, Gujarat Sahitya Parishad, Parishad Series 1, Bharatiya Vidya Bhavan, Ahmedabad.

Schoff, W.H., 1974. *The Periplus of the Erythraean Sea: Travel and Trade in the Indian Ocean*, Delhi.

Segal, D. and Carmi, I., 1996, 'Rehovot Radiocarbon Date List No. V', *Antiqot*, XXIX: 79-106.

Shastree, K.K., 1988. 'Where was Dwarka?', *Marine Archaeology of Indian Ocean Countries*, S.R. Rao (ed.), NIO, Goa, p. 25.

Shinde, V.S. and E. Thomus, 1993. 'A Unique Harappan Copper Fish-hook from Padri, Gujarat', *ME*, XVIII:145-147.

Somayajulu, B.L.K., W.S. Boecker, and J. Goddard, 1985. 'Dating Indian Corals by U-Th Decay Series Method', *QR*, 24: 235-239.

Sonawane, V.H., P. Ajithprasad, K.K. Bhan, K. Krishnan, S. Prathapachandran, Abhijit Majumdar, A.K. Patel, and J. Menon, 2003. 'Excavations at Bagasra-1996-2003: A Preliminary Report', *ME*, XXVIII(2):2150.

Soundara Rajan, K.V., 1994. *Kaveripattinam Excavations 1963-73: A Port City on the Tamilnadu Coast*, Archaeological Survey of India, New Delhi.

Souter, C., 1998. 'Stone Anchors Near Black Fort, Galle, Sri Lanka', *IJNA*, 27.4:331-342.

Sridharan, K., 1982. *Maritime History of India*, Publication Division, Government of India, New Delhi.

Srinivasan, S., 1984. 'Prabhasa or Prabhasa Mahatmya', in *Aspects of Indian History & Culture*, C.E. Ramachandran and K.V. Raman (eds.), New Delhi.

Srivastava, K.M., 1985. 'Bahrain-Indus Valley Connection', *Khaluj Times*, 23 August 1985.

Sundaresh and A.S. Gaur, 1998. 'Archaeology of Bet Dwarka', *ME*, XXII (2): 77-86.

Sundaresh, A.S. Gaur, P. Gudigar, Sila Tripati, K.H. Vora and S.N. Bandodker, 1999. 'Stone Anchors from Okhamandal Area, Gujarat Coast India', *IJNA*, 28.3:229-252.

Sundaresh, A.S. Gaur, Sila Tripati, P. Gudigar and S.N. Bandodkar, 2002. 'Stone Anchors from Bet Dwarka Island', Gujarat Coast, *BAIMA*, 26:43-50.

Sundaresh, A.S. Gaur, K.H. Vora, Sila Tripati and S.N. Bandodkar, 2005. 'Some New Thoughts on the Discovery of Ancient Dwarka', in *Revealing India's Past – Recent Trends in Art and Archaeology*, R.K. Sharma and Devendra Handa (eds.), Aryan Books International, New Delhi, pp. 22-32.

Technical Report, 1985. *Oceanographic Studies Off Sutrapada*, NIO, R.C. Bombay, pp. 7-8.

Technical Report, 1992. *Marine Archaeological Exploration off Dwarka and Somnath Waters*, NIO, Goa, pp. 6-7.

Technical Report, 2000. *Marine Archaeological Explorations off Dwarka, Bet Dwarka and Somnath, Phase-II*, NIO, Goa.

Technical Report, 2001. *Marine Archaeological Explorations off Dwarka, Bet Dwarka and Somnath, Phase-II*, NIO, Goa.

Taylor, P., 1976, *Marine Archaeology*, London, p. 192.

Thapar, B.K., 1991. 'Recent Archaeological Discoveries: Problem and Prospects', in *Cultural Heritage of Indian Village*, B. Durrans and T.R. Blurton (eds.), British Museum Occasional Papers, London, No. 47, pp. 1-9.

Tripathi, A., 1996. 'Dvaraka in Literature and Archaeology', *Man and Environment*, XXI (2):49-58.

Tripathi, S. and A.S. Gaur, 1997. 'Stone Anchors from Sindhudurg-West Coast of India', *IJNA* 26.1:51-57.

Tripathi, S., M.K. Saxena, Sundaresh, P. Gudigar and S.N. Bandodker, 1998. 'Marine Archaeological Exploration and Excavation of Vijaydurg – A Naval Base of the Maratha Period, Maharashtra, on the West Coast of India', *IJNA*, 27.1:51-57.

Tripathi, S., A.S. Gaur, Sundaresh, P. Gudigar and Bandodker, 1998. 'Historical Stone Anchors from Vijaydurg, Maharashtra, West Coast of India', *BAIMA*, 22 (1&2):1-8.

Tripathi, S., A.S. Gaur, Sundaresh and S.N. Bandodkar, 2001. 'Shipwreck Exploration in Sunchi Reef in Goa Waters, West Coast of India', *World Archaeology*, 32.3:357-69.

Tripathi, S., A.S. Gaur, Sundaresh, 2003. 'Anchors from Goa Waters, Central West Coast of India: Remains of Goa's Overseas Trade Contacts with Arabian Countries and Portugal', *BAIMA*, 27:97-106.

Tripathi, S., A. Manikfan and M. Mohamed, 2005. 'An Indo-Arabia Type of Stone Anchor from Kannur, Kerala, West Coast of India', *IJNA*, 34(1):131-137.

Vats, M.S., 1975. *Excavations at Harappa*, Bhartiya Publishing House, Varanasi, pp. 107, 332.

Vishnu Puran, 5.23.13: (Hindi Translation) Geeta Press, Gorakhpur.

Vora, K.H., D.K. Naik, P. Ganeshan and C.N. Moraes, 1991. 'Offshore Extension of Gomati River, Dwarka', *JMA*, 2:32-38.

Vora, K.H., A.S. Gaur, David Price and Sundaresh, 2002. 'Cultural Sequence of Bet Dwarka Island Based on Thermolumincence Dating', *CS*, 82.11:1351-56.

Vora, K.H., A.S. Gaur, Sundaresh and S. Tripati, 2006. 'Archaeological Sites as Indicators of Ancient Shorelines', *GMAI*, Proceedings of 7th Marine Archaeology Conference, A.S. Gaur and K.H. Vora (eds.), Society for Marine Archaeology, NIO, Goa, pp. 82-86.

Whitehouse, D., 1970. 'Excavation at Siraf: Third Interim Report', *Iran*, VIII:1-17.

Index